Biomimicry Materials and Applications

Scrivener Publishing
100 Cummings Center, Suite 541J
Beverly, MA 01915-6106

Publishers at Scrivener
Martin Scrivener (martin@scrivenerpublishing.com)
Phillip Carmical (pcarmical@scrivenerpublishing.com)

Biomimicry Materials and Applications

Edited by

Inamuddin

Department of Applied Chemistry, Aligarh Muslim University, India

Tariq Altalhi

Department of Chemistry, College of Science, Taif University, Saudi Arabia

and

Ashjan Alrogi

Hera General Hospital Makkah, Saudi Arabia

Scrivener Publishing

WILEY

This edition first published 2023 by John Wiley & Sons, Inc., 111 River Street, Hoboken, NJ 07030, USA and Scrivener Publishing LLC, 100 Cummings Center, Suite 541J, Beverly, MA 01915, USA

For more information about Scrivener publications please visit www.scrivenerpublishing.com.

Wiley Global Headquarters
111 River Street, Hoboken, NJ 07030, USA

For details of our global editorial offices, customer services, and more information about Wiley products visit us at www.wiley.com.

Library of Congress Cataloging-in-Publication Data

ISBN 978-1-394-16621-3

Cover image: Pixabay.Com
Cover design by Russell Richardson

Set in size of 11pt and Minion Pro by Manila Typesetting Company, Makati, Philippines

Printed in the USA

Contents

Preface

Biomimicry has been observed in the fabric industry since 4000 BC in China, but the recent works were started in 1950 by Otto Schmitt, a biophysicist developing the concept of *Biomimetics*. The practical applications of biomimetic materials are everywhere, from biotechnology to medicine to LED and beyond. These materials are creating a revolution and are the future of commercial work in nearly all fields.

Biomimetic materials are basically synthetic materials or man-made materials which can mimic or copy the properties of natural materials. Scientists have created a revolution by mimicking natural polymers through semi-synthetic or fully synthetic methods. There are different methods to mimic a material, such as copying form and shape, copying the process, and finally mimicking at an ecosystem level.

This book comprises a detailed description of the materials used to synthesize and form biomimetic materials. It provides the materials in a way that will be far more conventional and easier to understand. The authors have compiled the book so that it can be used in all areas of research, and it shows the properties, preparations, and applications of *biomimetic materials* currently being used.

Chapter 1 discusses the various types of animal eyes, biologically inspired optical systems, and various optical features in nature. The systems covered in this chapter have the potential for biomimicry.

Chapter 2 redefines cell and material interfaces. Host-microbe interactions, mimicking of extracellular matrix, and single-cell transcriptomics are explored, with the aid of colorful diagrams. Moreover, surface chemistry and topography are expounded by exemplifying cutting-edge discoveries. Additionally, for a deeper look, tumor angiogenesis, and vasculogenic mimicry were also elucidated.

Chapter 3 provides an overview of bacteriocins and their applications. Lactic acid (LA) producing bacteria, which are classified under the category of GRAS, generate a potential molecule known as bacteriocin, and this chapter discusses various classes of bacteriocins. The mechanisms of action on harmful bacteria are elaborated upon, as are various applications of LA-mediated bacteriocins in therapeutics and food industries.

Chapter 4 provides information on the properties of the keen biomedically applicable substrate, polydopamine (PDA). To give a complete picture, this chapter discusses dentistry specific, multi-purpose PDA coatings and nanostructured PDA materials, in addition to the structural properties. Overall, PDA applicability is highlighted.

Chapter 5 details various applications of electroactive polymer actuators in day-to-day activity. Chronological evaluation and their different groups is discussed in detail. This chapter covers some important engineering applications such as soft robots, underwater, aerospace, and energy harvesting. Additionally, alternative applications including wearable electronics, smart windows, and shape memory polymers are also covered.

Chapter 7 emphasizes the utility of electroactive polymers (EAPs) in actuator-based, refreshable braille displays through tactual means for the blind using cutting-edge digital technology, and how the flexibility of EAPs makes them superior to other actuation technologies based on their strain, elastic modulus, and ease of processing.

Chapter 8 deals with conceiving bio-inspired materials starting from natural structures, such as nacre, spider silk, and the gecko's foot, to possibly improve engineered materials, by coexisting toughness and strength, stress-modified performance, and substrate-independent reversible adhesion.

Chapter 9 discusses various biomimicry techniques for efficient management of unhealthy plants during pre- and post-harvest operations. The role of robotic techniques for on-the-field plant disease detection is discussed, as is the application of a support vector machine for the management of Gray-mold and Sclerotium rots in vegetables after harvest.

Chapter 10 discusses the role of biomimetic materials in textile industries. This chapter teaches the reader about the significance and advanced features of biomimicry fabrics such as algae fiber and mushroom leather. In addition, the future scope and challenges of fabric mimics are also discussed.

Our thanks go to Wiley and Scrivener Publishing for their continuous support and guidance.

Inamuddin
Department of Applied Chemistry, Zakir Husain College of Engineering and Technology, Aligarh Muslim University, Aligarh, India
Tariq Altalhi
Department of Chemistry, College of Science, Taif University, Taif, Saudi Arabia
Ashjan Alrogi
Hera General Hospital Makkah, Saudi Arabia

1

Biomimetic Optics

Priya Karmakar, Kripasindhu Karmakar, Sk. Mehebub Rahaman, Sandip Kundu, Subhendu Dhibar, Ujjwal Mandal* and Bidyut Saha†

Homogeneous Catalysis Laboratory, Department of Chemistry, The University of Burdwan, Burdwan, West Bengal, India

Abstract
The application of natural-inspired ideas and concepts in the development of technology solutions is termed as biomimetics. "Learning from nature" is a shorthand, i.e., nickname, for biomimetic design, which translates nature's superior designs to technological applications. Optical biomimetic blooms as a result of a recent rush of research into biological optical structures, resulting in a range of bioinspired photosensitive resources with improved visual characteristics facilitated via microstructures inherited from nature. The development of sensors and various optical devices, as well as inventive elucidation to pictorial challenges and the construction of smart systems, are all based on natural discoveries. The various types of animal eyes, biologically inspired optical systems, and various optical features in nature are all discussed in this article, which have the potential for biomimicry.

Keywords: Biomimetic, bioinspiration, visual systems, biomimicry

1.1 Introduction

When it comes to the advancement of optical science and technology throughout the years and centuries, which has resulted in novel optical working ideologies and components, it is common to find that the fundamental physical notions had already established in nature millions of years ago. Optical components, especially those that are very small, are often used in advanced visual communication and various direction-finding

**Corresponding author*: bsaha@chem.buruniv.ac.in
†*Corresponding author*: umandal@chem.buruniv.ac.in

Inamuddin, Tariq Altalhi and Ashjan Alrogi (eds.) Biomimicry Materials and Applications, (1–34)

devices, radars, therapeutic medical equipments, etc. In modern times, small-scale independent optical structures received a lot of attention, both for commercial and military purposes. The capability to reconfigure presentation in real time to perform any job, as well as quick flexibility to ecological changes via similarity with readily reprogrammable microelectronic devices and the ability to interact and communicate with microelectronic systems, are all essential requirements for these miniature optical systems. Traditional adaptive optical systems, on the other hand, are not yet as versatile, dependable, or transportable as their electronic equivalents. The mechanical technique, which has historically been employed to tune the performance of optical systems, is being phased out. Low moving component dependability because of the higher friction of surface on the very small scale, delayed reply, excessive use of power, and unaffordable price are some of the drawbacks.

Bionic ideas have now invaded a wide range of technical sectors. For one thing, bionics offers a vast array of systematically answers derived from the natural process of evolution (principles of first-kind evolution. Alternative optical devices that mimic the functions of living species' vision organs, such as vertebrate eyes (the human eye) [1–4] and insect compound eyes [5–7], are the best solution to the challenges mentioned above. Biological vision systems have evolved to be small, anatomically basic, and have a widespread range of adjustable functionality, as well as the capacity to adapt to changing environments through a variety of processes and reaction [8]. Owing to the opportunity of central length variation by lens redesigning, it is possible to produce eyelike single microlenses by simulating the human eye's adaptation function, by which we can substitute a sophisticated solid-state lens system [9, 10]. Optical components that are encouraged by the insect's compound eye can be used to make perfect optical sensors [7]. The following are a few instances of biomimetic principles in the optical realm:

(i) Scallop eyes include catadioptric systems, which combine imaging mirrors with traditional (refractive) lenses to reduce chromatic deviations in imaging structures for example optical microscope, photo lenses, or high-end lithography optics for the semiconductor manufacturing sector [11].
(ii) The dazzling cover of extreme-UV optics familiarized modern generation Lithigraphy using a similar physical process as the covering of facets in the cornea of a horsefly's eye in the semiconductor industry at 13 nm [12].
(iii) Having a photonic bandgap and gradually shifting in wave length range, photonic crystals, which are three-dimensional

structures with consecutive low and high refractive indices, signify that a fixed wave length value cannot propagate. Photonic-crystal fibers have been observed in nature, such as in an oceanic mouse, a naval maggot that looks like a glittering mouse. Photonic nanostructures also produce colors on butterfly wings and beetle shells [13].

(iv) The notion of flexible refractive surfaces was extended to liquid lenses that allow artificial optical systems to focus and zoom [14, 15]. This is similar to the malleable eye lenses used by reptiles, birds, and mammals to adjust their eyes. However, creating a comprehensive technical duplicate of the human eye—replete with a muscular-like actuation—is still a huge technological hurdle.

Bionics has progressed to a new degree of technical significance in the twenty-first century. Bionics not only offers a series of self-optimized solutions but also gives us very effective routes to get there [16].

Some examples of biomimetic concepts are depicted in Figure 1.1. They demonstrate how natural design combined with artificial materials has immensely helped humanity [17].

Because of its abundance of optical structures with optimum properties, nature outperforms humans. Numerous microstructures, especially based on multilayer, photonic crystal, and nonlinear network [20], demonstrate advanced capabilities in physical light manipulation. Modifications to the fundamental structures result in a wide range of alternative structures with the addition of other optical processes, resulting in sophisticated optical effects, which are particularly frequent in multilayer-based optical structures. These fascinating optical phenomena are mostly based on five basic principles of optics: (i) interference in a thin-film, (ii) a deflection grating effect, (iii) multilayer interfering, (iv) photonic crystals, and (v) light sprinkling [21]. The appreciation of humans has been sparked by a variety of perfected optical designs seen in nature. In optical engineering, there are additional benefits for humans. A variety of optical effects using microstructures which have low refractive indexes has been accomplished by the nature. The prevalent component in invertebrate reflectors, guanine, has a refractive index of 1.83, which makes it a high refractive index biological material [22–24]. Other features of materials, such as liquid crystals and photosensitive polymers, in addition to refractive index contrast, considerably enable optical engineering. "Learning from nature" is crucial and significant for designing optical structures, such as reflectors with multiple layers, crystal fibers sensitive toward photon, and light abstraction systems [25].

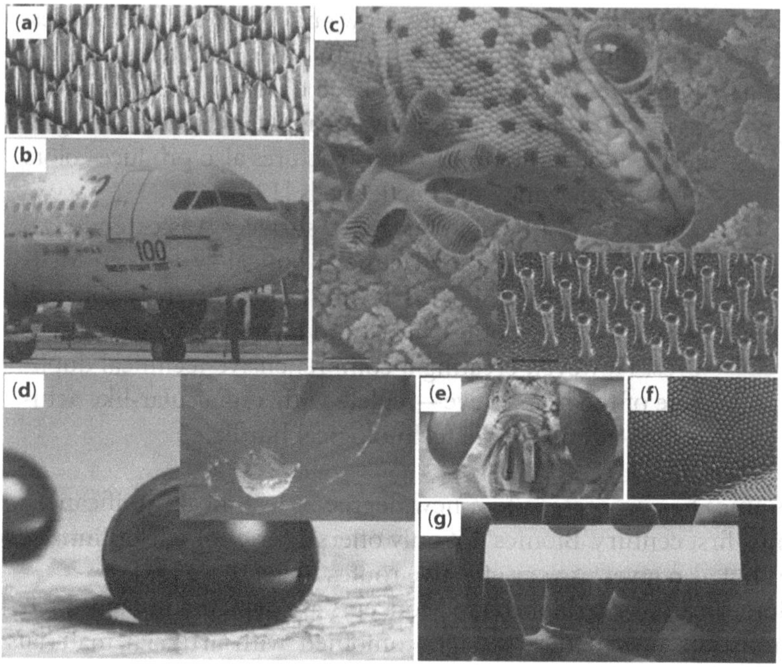

Figure 1.1 Distinctive biomimetic substances. Image of smaller pieces cut from a full rack of skin of shark (a) Coating of microscopic plastic layer on aeroplane (b) Setal structures on foot (background) of the lizard G. gecko (c) are used for attachment. Superhydrophobicity of "Lotus Spray" is demonstrated by water droplets that resemble non touching nature of water in the leaf of lotus. (d) Antireflection effects are visible in the Calliphora sp. (e) due to subwavelength features on the ommatidium's surface (f). The plane of the glass (g) having antireflection effect. [Scale bars (inset c) 2 μm, (background image c) 10 μm] [18, 19].

1.2 What is Biomimicry?

From demonstrating aeroplanes which used to be inspired by flying of bird, to building robots and technological advancement in the present day, we have studied the supremacy of nature for a thousand years, determining to emulate its excellently crafted systems. During thc 1950s, an eminent bio-physicist named Otto Herbert Schmitt coined the word "biomimetics." The meaning of the word is "to replicate life and nature," and in 1960, at a symposium at Wright-Patterson Air Force Base in Dayton, Ohio, he coined the term "bionics." Biomimicry, also known as bioinspiration, is the practice of employing nature's design principles to develop incredibly intricate and sophisticated manufacturing models at numerous length scales and then applying that knowledge to tackle humanity's most pressing problems [26, 27]. This field is becoming more closely associated with developing scientific and engineering disciplines. This branch of research has advanced significantly in recent decades through the study of the earth's flora and fauna, with an emphasis on insects, marine organisms, and people. It has improved greatly in recent decades, and now it has vast applications in sectors such as architecture, material science, engineering, medicine, and robotics [28]. Learning from nature's "solution handbook" and obtaining inspiration or even directly using design concepts to solve several major global concerns such as public health, food, national security, water, and renewable energy will be a very effective strategy.

The essence of biomimetics is the understanding and adopting nature's techniques for the applications in real life scenarios. Biomimetics is divided into two categories by Gebeshuber and Drack: analogy biomimetics and induction biomimetics [29]. The former begins with technical issues and seeks answers from equivalent circumstances in nature, such as attaching winglets to aeroplanes. Biomimetics through induction has no aim of being used at first, but instead begins with biological events and considers how to apply them to the creation of technology, such as the discovery of Velcro from bursts of air. Understanding nature's problem-solving tactics is not as simple as one may imagine. Biomimicry is defined by two fundamental elements: observation and creativity. It is not merely replicating nature, but rather absorbing clues from it to produce an efficient multifunctional system. The evolution of the eye and brain, as well as how they function in various organisms, has long fascinated researchers.

Singamaneni and coworkers' overview summarizes the optical ingredients and phenomena in nature that are essentially the result of evolution's arms race, natural selection. The sun's light, for example, has affected the evolution of natural selection on the planet. As a result of their capacity to see and sense, certain animals have evolved a spectacular color palette as a result of breeding preferences, pictorial communication, and signalling. Aside from nature's amazing color contrast, there is a huge variety of photosensitive structures that have resulted to more advanced visual indications including ultra violet and infrared vision, polarized vision, and also night vision. Nature exhibits remarkable optical phenomena such as reflection and antireflection, tailored absorption and transparency, interfering, light harvesting, scattering, diffraction, waveguiding, and bioluminescence, lensing which are all responsible for the exceptional optical properties of biological structures and materials. Bioinspired materials and approaches have unquestionably expanded the scope of various optical devices, sensors, and material development.

1.3 Step-by-Step Approach for Designing Biomimetic Optical Materials From Bioorganisms

Biomimetic design is the process of transferring great design principles of nature to technological utilizations. Numerous strategies have been used to analyze and define the physical characteristics of biological systems. The findings, which are typically gathered utilizing a variety of microscopic, optical interrogation, and modelling approaches, were extensive. Analysis of optical microstructures in nature is the initial outset in fabricating biomimetic optical materials with structure design inspired by nature. Then, in order to identify the underlying optical processes, optical characterization is performed [30]. The last phase incorporates nature's design into man-made materials via meticulously planned production techniques.

1.3.1 Optical Structure Analysis in Biology

The characterization of optical microstructures in nature is the initial stage in optical biomimetics. Pigmentary and structural color effects are the main two types of color effects in biology. Pigmentary color results from particular light absorption by pigment electrons, whereas structural color results from selected light reflection caused by interference, diffraction, and scattering. Structural color refers to the optical effects. When there is no optical absorption, a dielectric material's refractive index (RI) is solely real, but when light is absorbed, it contains both real and imaginary components [31]. The optical microstructure in the sequence of wavelengths of light is the source of structural color. Aside from structural color, optical structures may modify light flow to create other optical phenomena including antireflection, polarization, and light focusing. Optical microstructures are dominated by butterflies, beetles, and birds and the structural origin of optical phenomena has been recognized.

Two sheets of chitinous scales are patterned laterally across the dorsal and ventral flaps along both sides of the butterfly's wings. The basal scales are placed immediately on pinnacle of the wing membrane, like roofs on a house. This is depicted in Figures 1.2a and b. Within the wing membrane, each scale is a flattened cuticle protrusion from a single epidermal cell. Scale dimensions are approximately 75 by 200 μm [32]. As shown in Figure 1.2d, the derivation of each structural component, namely the ridge, microrib, and internal structure, results in a variety of microstructures. Barbs with

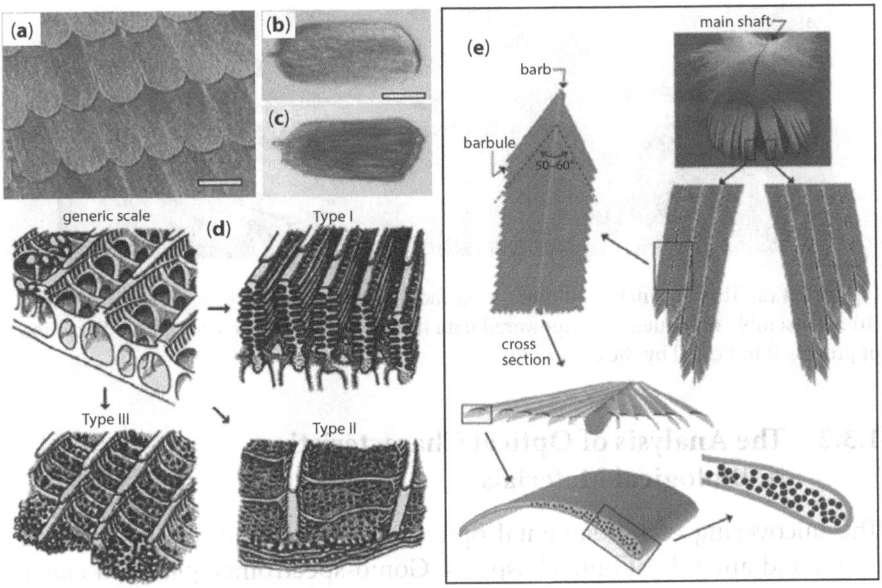

Figure 1.2 (a) Perceptible picture of M. rhetenor's scales of wings. (b) Dorsal areas and (c) ventral areas M. rhetenor scale. (d) Geometrical categorization of shimmering Lepidoptera lamella. The microstructures seen on the moths and the butterflies are depicted in this schematic picture. (e) Schematic representation of neck feather of the rock dove. Detailed features inside barbules and cross-sectional structures are sketched [Scales of wings 100 μm, dorsal and ventral areas 25 μm] [33–35].

barbules (has optical multilayer microstructures) are embedded in the primary shaft of bird feathers is shown in Figure 1.2e.

Some very important methods to mention are SEM, TEM, AFM, FIBSEM, etc. FIBSEM suppliments the 2D viewing limitation of traditional SEM. Figure 1.3a depicts a punch a three-cornered notch into a butterfly with FIBSEM, with a side length of roughly 10 μm, allowing the internal structure to be seen straight away. Sequential FIBSEM, in combination with computational reconstruction, reveals a diamond structure at beetle size (Figure 1.3b) [37]. These computer-assisted microscopic procedures are nonetheless time-consuming and challenging, particularly when it comes to developing a section. 3D photonic crystal structures at single scales without the requirement for pre-treatment can be characterized by small-angle X-ray scattering (SAXS). The gyroid structure of butterfly scales has been studied using SAXS [38]. SAXS is useful for characterization of 3D photonic crystal structures on single scales without the requirement to pre-treat the scales.

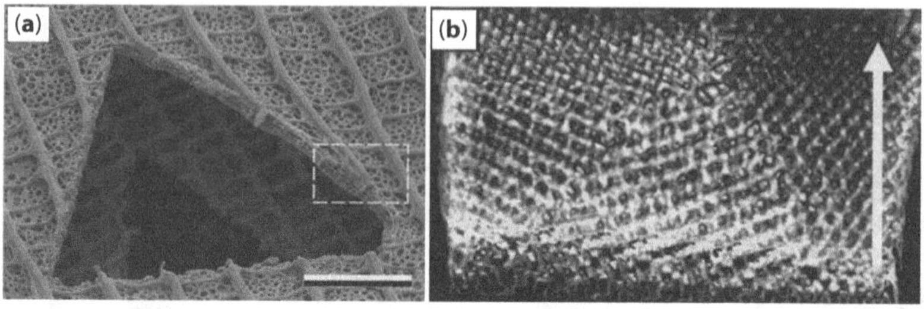

Figure 1.3 (a) Three-cornered pit into the surface of pinion of grass blue (Zizina labradus). (b) 3D assembly of sequential empowered data from the FIBSEM by a computer. Direction of process is indicated by the arrow [30].

1.3.2 The Analysis of Optical Characteristics in Biological Materials

The uncovering of fundamental optical processes requires the interpretation and analysis of optical aspects. Gonio-spectrometry is used extensively along with the spatial scatterrometry (Figure 1.4) in this purpose. Figure 1.4a schematically depicts a simple and direct approach for detecting light scattering from biological materials. It is a technique that has been applied for a variety of animal samples of varied sizes. In nature, optical structures are frequently made up of microscopic components such as scales and barbules, so to investigate the optical process, it is necessary to

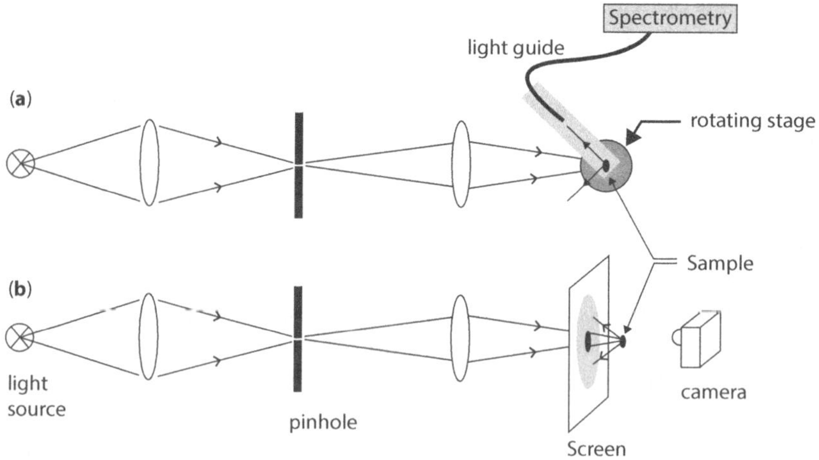

Figure 1.4 (a) This graphic depicts the equipment to determine the angular distribution of scattered photon, (b) Schematic representation of perceptual dispensation of light by butterfly scale [36].

optically characterize each small skeletal component. The metric settings, particularly the light source and lens, must be precisely prepared because of the very small size of optical microstructures. Researchers employed a gonio-spectrometric system that included a beam of laser as the source of light, micromanipulator along with a lens which is achromatic to characterize a single butterfly scale [39]. Figure 1.4b depicts a simple and effective way for viewing a little to medium-sized object, like a bird feather barbule or a butterfly scale disperses spatial light.

Calculation prediction is commonly used to aid in the explanation of complicated optical effects in nature. On lamellar and crystal-like arrays, traditional approaches, such as Bragg's law computer modelling and thin-film optics are used. Most importantly, selected light source for spectrometric research is sufficiently wide and have adequate intensity over the full wavelength range of interest. After an appropriate warm-up period, this intensity should either not change with time or be compensated for in the data processing [40]. The 2-D Fourier analysis may be used to estimate a biological nanostructure's back scattering reflectance spectrum by varying in refractive index. Scientific study on Morpho butterflies has advanced quickly during the 1990s, with a variety of extensive observations and complex computations and this approach may be used to directly compute the electromagnetic field surrounding the optical microstructure as well as the distant field [41–43].

1.3.3 Optical Biomimetic Materials Fabrication Strategies

Nature is a great resource of creativity for optical structure design and fabrication for a number of applications. A butterfly's structural colors have been studied [44–46]. Optical biomimetic refers to the creation of materials that mimic natural optical microstructures. A number of microfabrication processes have been developed, each with its own set of properties. Biomorphic mineralization is a simple, low-cost process that can replicate complicated microstructures seen in nature. Precursors (vapor, liquid, solid) are deposited or penetrated into biostructures as templates. Synthetic reproduction of clearly valuable structures of biology using an easy casting technique is utilized to bring a few biological structures' superior qualities to man-made materials [47, 48]. Biomorphic mineralization is a typical method for producing optical biomimetic materials. Micropatterns that resemble natural microstructures are frequently created using electron beam lithography. Nanoimprinting and soft lithography are low-cost, effiecient-throughput production technologies that may be used to make optical biomimetic materials with flat and basic 3D microstructures [49]. Additive manufacturing which is also known as 3D printing has unfolded new possibilities for altering along with replicating objects that are inherently multiscale, multimaterial, and multifunctional. It might be a useful technique for creating speculative strategies to better explore biological process and its interplay with environmental restrictions in manmade systems, as well as for overcoming the manufacturing challenges posed by biomimicry research [50].

1.4 Biological Visual Systems—Animal and Human

The first scientific theories on vision emerged in ancient Greeks time. They defined sight as rudimentary fire erupting from the eye. In its most basic form, sight requires that rays of light be concentrated on a detector which detects light. Visual systems have evolved in a converging and progressive manner, with numerous species achieving identical functional goals via diverse routes.

Animal eyes have developed from rudimentary structures that sense light, such as those found in flatworms, to complicated systems that can generate images, such as those seen in insects and humans. The structural intricacy of biological eyes is astounding, with components ranging in size from nanometres to centimetres. Correlative biological investigations reveal the nature of other creatures' visual systems differ from ours. There are a different kind of eye forms and structures which support the animal's path of existence, on the basis of whether it is hunted or hunter [51]. Nature has revealed a total of ten generalized optical systems, each with its unique set of variants. Most important types of animals eyes are: (1) camera-type eyes or single-chambered simple eyes and (2) compound eyes.

1.4.1 Simple Eyes

Simple eyes contain only one element which focuses the light onto a neurological component [52]. This optical device in the animal kingdom might be a mirror, a normal slit, or, more typically, a lens. Nautilus's eyes, which are like pinhole, only allow the light to pass through to its retina [52]. This eye structure simply facilitates the organism in detecting light and its direction.

Simple eyes which are camera-type are very well known among this category, as they resemble a camera. The principal parts in this type of eye are lens and retina. Such an eye type may be found in vertebrates, aquatic species such as cephalopods and mollusks, and arthropod arachnids (Figure 1.5) [51]. The refractive index and lens shape alter according to the medium (water or air) of outside of eye.

The most well-known biological optical system is the human eye. It is called a camera eye since it just has one lens for focusing images onto the retina. Four tissues which are responsible for refraction, in the eye of human, work together to generate a picture on the retina. The cornea has a fixed focal distance and controls the eye's ability to focus. If the intensity of light is too intense, the pupil closes to minimize the light flow and protect the delicate retina. In strong light, the diameter of pupil ranges from 2 to 4 mm, whereas in dark it ranges from 4 to 8 mm. The clear bi-convex structure of the crystalline lens, which is also flexible, enables the eye to

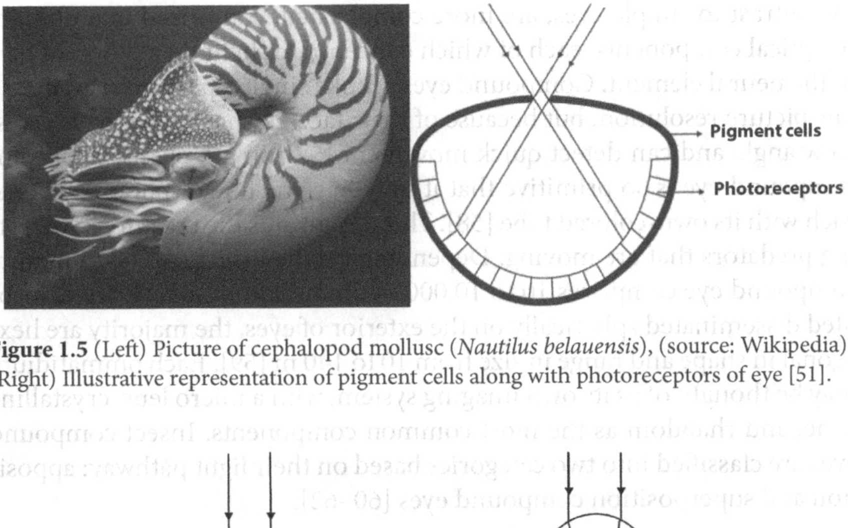

Figure 1.5 (Left) Picture of cephalopod mollusc (*Nautilus belauensis*), (source: Wikipedia). (Right) Illustrative representation of pigment cells along with photoreceptors of eye [51].

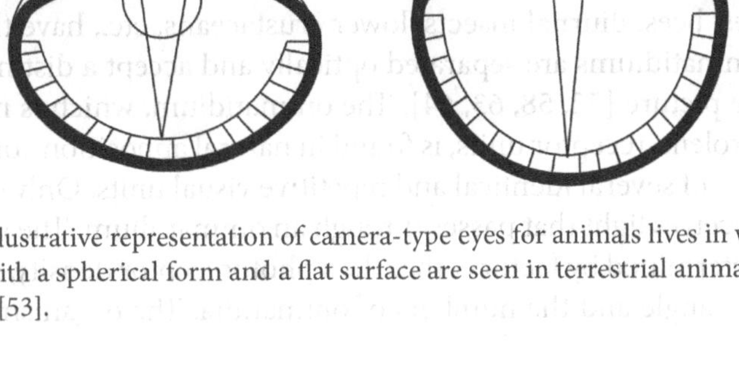

Figure 1.6 Illustrative representation of camera-type eyes for animals lives in water (left) and lenses with a spherical form and a flat surface are seen in terrestrial animals (right), respectively [53].

change the distance of focus in between the lens and the object (Figure 1.6) [53]. The term "accommodation" refers to this process. Aqueous humor separates the aspherical cornea from the biconvex gradient index (GRIN) lens in humans [54]. Ciliary muscle contractions govern alterations in the radii of curvature of the lens via zonule tissue, and as a result, focal length changes [55]. Rods and cones (two very important receptors of photons) and connecting neurons are responsible for the transfer of information to the brain via the retina. The capacity to follow moving objects and stabilize the picture of the item is also an essential adaptive function of the human vision system [56].

1.4.2 Compound Eyes

Optical engineers have been focusing on the compound eye in order to emulate the architecture in imaging systems and sensors. Compound eyes, in contrast to simple eyes, are more complicated. comprised of a number of optical components, each of which concentrates light on a different part of the neural element. Compound eyes, unlike single-aperture eyes, have a low picture resolution, but because of their faceted shape, they have a vast view angle and can detect quick movement [57]. In ark clams, the proto-compound eye is so primitive that it only has one line of photoreceptors, each with its own-colored tube [58]. This design aids the creature in detecting predators that are moving. Depending on the insect species, a natural compound eye comprises 10 to 10,000 of ommatidium, which are distributed disseminated spherically on the exterior of eyes, the majority are hexagonal in shape and range in size from 10 to 140 m [59]. Each ommatidium may be thought of as its own imaging system, with a micro lens, crystalline cone, and rhabdom as the most common components. Insect compound eyes are classified into two categories based on their light pathway: apposition and superposition compound eyes [60–62].

1.4.2.1 Appositional Compound Eyes

Dragonflies, bees, diurnal insects, lower crustaceans, etc., have this kind of eye. All ommatidiums are separated optically and accept a distinct component of the picture [52, 58, 63, 64]. The ommatidium, which is made up of single microlens-receptor units, is found in natural apposition complex eyes and consists of several identical and repetitive visual units. Only one photoreceptor receives light that passes through an ommatidium. The visual resolution is determined by factors other than photoreceptor density, such as the ommatidial angle and the numbers of ommatidia. The organism's ultimate picture is vertical, and it is made by sewing together all of the ommatidia's images. The pigment cell is inserted to the intermediate space of neighboring ommatidia to reduce optical crosstalk [65]. As, the apposition eyes have tiny lenses with a diameter of 10µm, they provide incorrect visual signals at low densities. As a result, they perform best at high highlight intensities. In bees, all ommatidia photoreceptors point to almost the similar kind of vision, whereas in houseflies and drosophila, every receptor has its own view which is distributed with nearby ommatidium receptors [51]. A similar neural bundle is created by all this-axons to receive information from seven neighboring ommatidia. This unusual neuronal architecture is known as neural superposition, and it is similar to the spatial summation mechanism of human rod photoreceptors, which boosts neuron sensitivity [52, 66]. To boost light sensitivity, the rhabdom in certain flies is split into several light guides. Individual light guiding signals are neutrally accumulated to generate a powerful response from the brain.

1.4.2.2 Superpositional Compound Eyes

Light from various aspects is routed to a general receptor to generate a picture of the object, demonstrating that superposition compound eyes are substantially more sensitive towards light than apposition eyes (Figure 1.7)

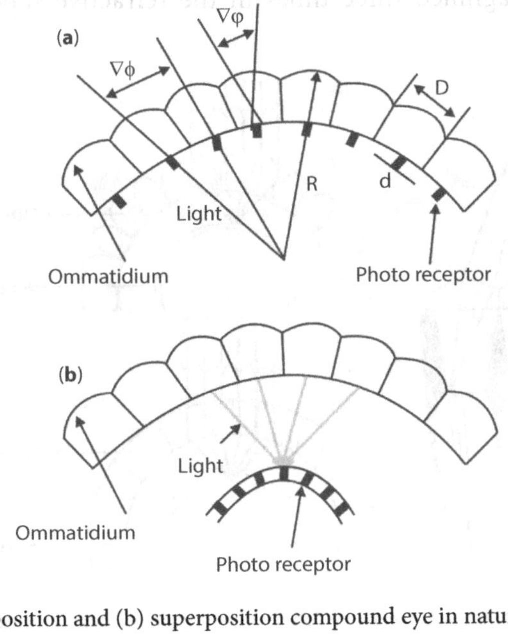

Figure 1.7 (a) Apposition and (b) superposition compound eye in nature [67].

[67]. Furthermore, the superposition compound eye has a 10 fold greater imaging resolution than the apposition compound eye [68]. In contraty to apposition eyes, superposition eyes' visual units do not operate independently because hundreds of neighboring facets work together to collect light, expanding the aperture. This genotype has been linked to deep-water crustaceans and nocturnal insects that live in low-light environments [62]. Refractive, reflective, and neural superposition are the three forms of super-positional compound eyes, as depicted in the diagram.

Optical superpositional compound eyes (refractive and reflective) are commonly found in moths, lobsters, nocturnal insects, and fireflies, and moths. By integrating optical input from neighboring ommatidia in a neural line, which is single and isolated from the optical units, the neural superpositional compound eyes increase light sensitivity more than the neural superpositional compound eyes. The superpositional eye contains a slender rod that not only functions as a light guide but also structurally separates the crystalline cone and the rhabdom. Furthermore, because of the lack of a shield consist of pigment, every rhabdom may collect light from its own ommatidium along with that of other nearby Ommatidia [52, 63]. In moths and many beetles, the refracting superpositional eye has a descent refractive index and this suggests that light bends occur within the lens (Figure 1.8(b)). Light is directed to the photoreceptor in refractive superposition eyes via the ommatidial lenses' refraction [51]. Photosensitivity is magnified three times in the refractive superposition eye,

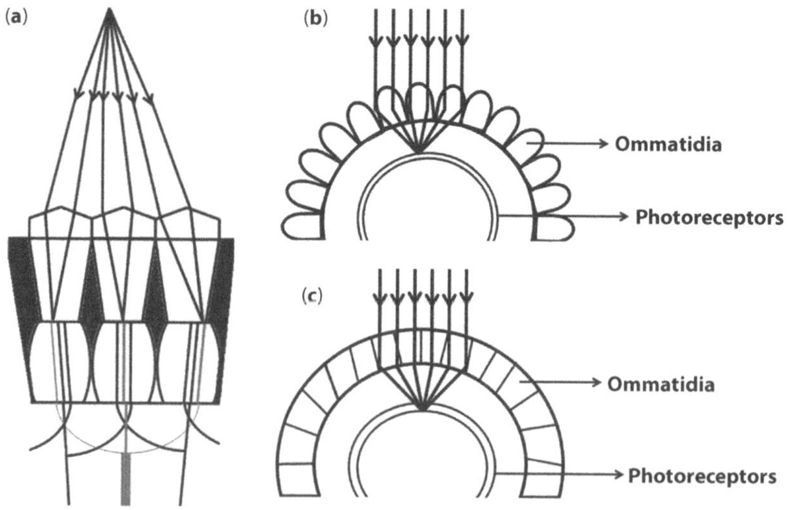

Figure 1.8 (a) Neural (b) refractive (c) reflecting superpositional compound eyes [51].

compared to ordinary eyes. The outside walls of the ommatidia of lobsters and shrimps seem silvered, so incoming light is mirrored on the wall's sides and reaches the receptors which absorbs and senses the light (Figure 1.8(c)). To remove scattered reflections from various angles, the ommatidia are square in form. Each ommatidium in the neural superposition eyes is made up of numerous rhabdoms that are overlaying this very same nerve (Figure 1.8(a)). Because of this result, incoming light from various lenses is gathered on various rhabdoms inside a single ommatidium. Because one optical nerve combines several rhabdoms, it can give improved resolution without raising the ommatidium.

Even though some organisms lack an eye, they are able to detect light. Chiton and brittle stars on their dorsal surfaces have crystals for fixed lenses, which help to focus light through neurons and makes sensitive toward photons [69, 70]. For example, Melanophila acuminate uses some special frameworks termed as "pit organs," to perceive the infrared part of electromagnetic radiation [71]. This sensibility aids them in sensing forest fires and laying their eggs on the charred wood.

1.5 The Eye's Optical and Neural Components

From the cornea to the percept, we will go over a few of the distinctive aspects of the eye's optical parts and neurological systems of eye. that may be found in nature, as well as bioinspired materials and technologies applied to each section of the eye.

1.5.1 Cornea

The cornea, or more often known as the optical window, has evolved into a substantial refracting element in vertebrates, which results almost 2/3 parts of the power of eye. Unlike the tiny eyes of Nautilus, it works as a protective barrier by protecting the eye's content from the surroundings [72]. As having a larger curvature radius at the perimeter than at the middle, the human cornea results in an aspheric design, which helps to eliminate aberration [73].

Cornea with a biomimetic structure

Cornea is the anterior layer of the human eye, and it is damaged easily by chemical damage as well as physical trauma, requiring wound care in the early phases and replacement of cornea in more dreadful situations. Despite

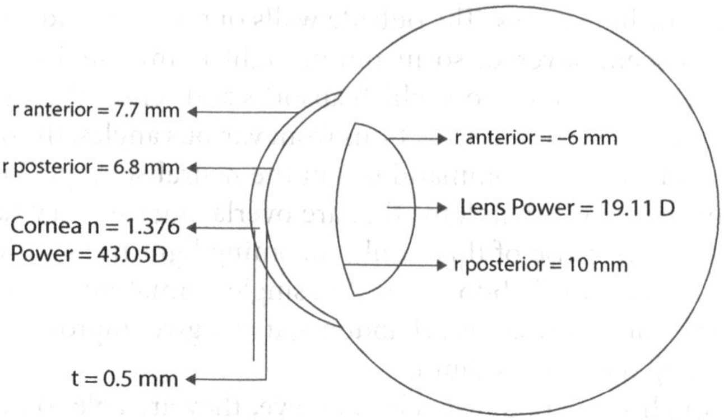

Figure 1.9 Diagrammatical representation of the eye. (n is the refractive index, r is the curvature radius, t is the diameter) [51].

worries about adulterant, the transmission of contagious infection, and the unsatisfied demand for donor tissue, studies have suggested at a variety of ways of developing biomimetic transplants of cornea (Figure 1.9) [51]. Electrospinning has sparked considerable interest in fabricating biomimetic nanofibrous wound dressings [74]. A group of Chinese researchers has invented a biomimetic nanofibrous membrane that is optically transparent and contains an internal layer of collagen–hyaluronate and a surface layer of chitosan [75]. Collagen type I, which makes up the majority of the protein in cornea, is non-toxic, biodegradable, and immunogenic. Collagen, being the main constituent of the extra cellular matrix (ECM), is well suited for the application in corneal scaffolding which is engineered with tissue. Similarly a polysaccharide HA, is required for corneal cell survival, as well as for corneal tissue repair. The goal of this research was to create a novel collagen or HA-based nanofibrous membrane (CBNM) as a treatment of corneal wound using electrospinning and chitosan (CS) modification. For example, nanofibers electro-spun were applied along with these materials on alkali-based eye injuries in rats, and it was shown to have much higher mechanical and biological efficacy than the typical human amniotic membrane.

1.5.2 Pupils

Many animals have more common slits or oval-shaped pupils, while humans have a circular pupil [76]. The shape of an organism's pupil is determined by animal's living environment as well as whether it is a predator or a prey species [77]. According to Banks *et al.* [77], the grade and clarity of picture created on the retina are influenced by the shape of pupil. When horses and goats lay their heads down to feed, their horizontally positioned pupils let them monitor for predators in a nearly 330^0 perspectives [77]. When a point source is photographed by an optical system, the result is frequently a blurry circle. Aspects such as aperture size and wavelength influence the size of PSF and MTF and the geometry and shape are determined pupil's shape. As a result of the disparity, the depth of field is astigmatic, meaning it is longer vertically than horizontally [77]. This implies that things beyond the focal distance will be blurred more horizontally than vertically, which may assist the animal in measuring the range of its prey. As the angle of incoming light grows, the circular diffraction pattern morphs into a horizontally expanding hyperbola, lowering the effective pupil area and contributing to diminished peripheral vision [78]. Slit pupils have somewhat superior contrast transfer function than circular pupils.

When a grating with a contrast of 100% and a cutoff frequency of 0.8 is photographed, with a circular pupil, the contrast is only 10%; however, with a slit pupil, the contrast is 20% [79]. To investigate the benefits of having different pupil shapes, a typical camera might be customized with different apertures.

1.5.3 Lens

In humans and other animals, the lens is primarily responsible for accommodation and aberration reduction. Various properties, functions, applications of lenses, are discussed here.

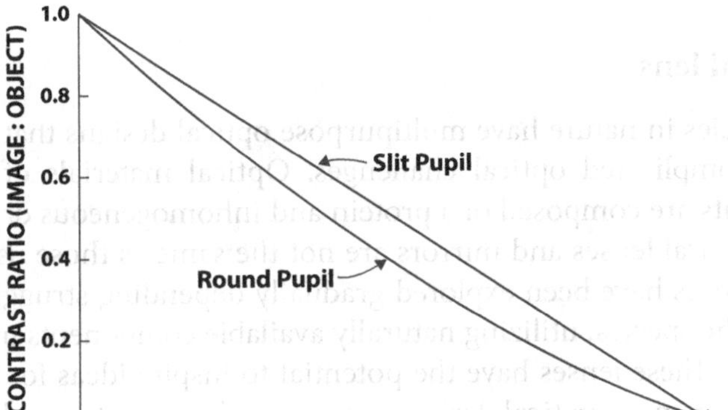

Figure 1.10 Graph of contrast ratio vs spatial frequency. *X*-axis is spatial frequency, *Y*-axis is contrast ratio of the object [51].

Accommodation, or the ability to see effectively at different ranges, is achieved by different kind of systems in various animals, wherein, Mammals modify their curvature of lens, so that it can focus on things in various distances (Figure 1.10) [51]. Snakes can change the placement of their lenses to move them nearer or in more distance from retina [52, 64]. The ciliary muscle contracts to regulates the curvature of human lens, allowing it to focus on things at various distances [54]. The pressure in the vitreous is so high in that case that it pushes the vitreous ahead. This mentioned process improves the eye's focusing ability [80]. Optics and haptics are two sections of an intraocular lens (IOL) that help to maintain the lens in its bag. CrystaLens is an accommodating intraocular lens (IOL) developed by Bausch and Lomb (NY). It functions by shifting the IOL to the front and backwards along the eye's axis [81].

Another significant word is aberrations, which occur when an imaging system's optical transfer function degrades, resulting in poor image quality. Biological lenses have evolved to minimize imperfections and give excellent retinal vision. The increase in lens diameter causes an increase in aberration, which helps to explain why arthropod ommatidia produce aberration-free images due to their tiny lens array. A spherical lens has the drawback of bending the periphery rays more, resulting in spherical aberration. On the other hand, spherical lenses having a gradient refractive index, lowers from middle to perimeter. As a result, all rays concentrate on a fixed end and shows the perspicuous picture [63]. The lens must be inhomogeneous when the focal length-to-radius ratio is less than Matthiessen's ratio value, 2.75 [63]. A similar graded index is seen in the human lenses, however in biconvex form that reduces spherical aberration [82].

Biological lens

Many species in nature have multipurpose optical designs that help them address complicated optical challenges. Optical materials of biological components are composed of a protein and inhomogeneous organic molecules. Natural lenses and mirrors are not the same as those used in labs. Natural lenses have been explored gradually depending struggle for existence on the species', utilizing naturally available components in their surroundings. These lenses have the potential to inspire ideas for improving existing man-made optical designs; one example is mentioned below.

The brittle star, Ophiocoma, is a marine species that displays luminescence by changing body color and through light-induced motion known as phototaxis [69]. Single calcite crystals are employed in the skeletal structure of brittle stars and are a constituent of sophisticated photosensory organs that could function analogous to compound eyes. Their dorsal arm plates have an array of double-lens-shaped microlenses that focus light onto photosensors (nerve bundles). Calcitic microlenses might form a specialized photoreceptor system with compound-eye properties owing to the unique focusing action and underlying neuronal sensors. Aizenberg *et al.* [69] used the photolithographic approach to investigate the lenses' focusing capabilities, constructing an man-made brittlestar lens arrangement and estimating average size of diaphragm, in which the diameter of the lense is L and amplification factor of light is E. When the values of L and E were found to be rather high, it was suggested that the lenses were adjusted for the spherical aberration. A brittlestar's body adjust their position in response to arm progression, enabling them to retrieve a substantial sense of visual data in regard to their surroundings. During the day, chromatophores around brittle stars' lenses stretch their processes across the lenses to control the amount of light which is entering through the lens, similar to how pupils restrict the amount of light. For this reason, when the animal is exposed to light, its body color appears to change. Brittlestars' use of calcite demonstrates organisms' astonishing capacity to optimize one material for several tasks via evolution, and presents fresh thoughts for the manufacture of "smart" substances [83].

1.5.4 Retina

The number and spacing of photoreceptors, as well as the nature of the optical picture, affect the perceived image quality [52]. Deinopis, an Australian spider has poor resolution but enough sensitivity to catch its prey at night. The retina's location in relation to the lens is also critical for obtaining a high-quality image. The retina is significantly closer to the lens in gastropod mollusks' tiny eyes and dorsal ocelli in insects' dorsal ocelli, allowing the creature to just perceive the quantity and quality of light from various angles, rather than resolving details of the objects [52]. Drosophila, a fruit fly, can only distinguish four objects separated by 0.5 minutes of arc, whereas humans can discern two. Three kinds of cone photoreceptors exist in the human retina, each of which is susceptible to covering sections of the visible spectrum. In low-light situations, however, they are practically color blind. Tarentola chazaliae, the nocturnal helmet gecko, bears massive poles with three distinct pigment molecules that are, respectively, UV, blue, and green wavelengths. Due to the presence of a multifocal visual system, they can compensate for chromatic aberration and their sensitivity for color is said to be 350 times that of humans. Light must penetrate through the whole thickness of the retina before it can be detected by photoreceptors. The optically inhomogeneous retina is subjected to a great quantity of light scattering as a result of this configuration. Despite the diversity of eye patterns found in nature, the underlying concept of picture production is universal. Brain knows how to interpret the image obtained in retina, even if upright and inverted, i.e., irrespective of the orientation of the image, since the images developed together with the eye [52].

1.6 Application of Biomimetic Optics

Thousands of years of natural "trials and mistakes" have resulted in a massive library of efficient answers to technical difficulties with biological species' existence. Combining natural design aided civilization, as proven via biomimetic models. Due to its abundance of optical structures with optimum properties, nature outperforms humans. Photonic crystal, irregular network, and multilayer microstructures demonstrate extensive light manipulation capabilities. Because of their superior optical system, most biomimetic vision systems are now ispired by optical imaging systems based on vertebrate principles and these have been used and widely deployed [84]. Medical equipments, consumer electronics, machine vision systems, smart cameras, and robots all benefit from optical systems that imitate eye functions. This optics has a number of benefits over traditional optical technologies, including better adaptability to changing circumstances and a wide range of functional properties in small packages.

1.6.1 Hybrid Optical Components are Meant to Resemble the Optical System of the Eye

In comparison to typical solid-state competitors, liquid-based optical components have the possibility for flexible adjustment of optical characteristics and compatibility, making them fascinating prospects for the formation of microscopic optical systems replicating the animal's eye reactions. Various physical characteristics of fluids, such as surface tension, mass diffusivity, fluidity, heat, and fluidity, make it feasible to swiftly and easily change interfacial geometry. These properties of fluid can also help to form smooth refractive surfaces on the molecular level, and by mixing liquid components it can also change the refractive index.

Systems containing hybrid optical configurations may mimic the core structure of eye response, the actuation principles and eye structures. The creation of biomorphic vision organs (artificial eyes) for use in vision systems for robots (mobile), industry, for ophthalmological study are one of the most important applications of such hybrid systems. The lens is generally the adjustable part of such systems, and the actuation idea is usually based on the concept of using artificial muscles to serve as analogues of the iris or ciliary eye muscles.

A bionic robot eye with an eyeball-like optical system [85]. In this eye model, the cornea and crystalline lens simulate accommodation. Shifting actuator changes the focal length, and because of the changes in pressure, distortion and dilation of the membrane takes place. According to the test findings, the suggested method at various distances is comparable to the human eye's accommodation characteristics [2]. Accommodation, or focal length adjustment, is accomplished by exerting pressure to the material across the edge boundary. This process results diameter and curvature change of the lens, similar to that of a crystalline lens. To produce the tension, an electric field or physically connected metal clamps are used so that it can control a dielectric elastomer actuator. The use of a dielectric elastomer results in a focal length adjustment range that is comparable to that of the eye, as well as a very short reaction time [2].

The example of a compact biomimetic imaging system, inspired by the human eye, which combines a liquid-based iris with an adjustable polymer lens, was presented in [86]. The device uses two separate actuation methods to show the pupillary light reaction and accommodation. An elastic plano-convex body connected to a motorized actuator was employed, similar to the crystalline lens. By replacing the motorized actuator, the eye-mimicking device can be improved with an artificial muscle that approximates a thermally sensitive liquid crystal elastomer (LCE). The adjustable device was fused with an image sensor chip and a set of solid-state lenses. The functional capabilities and visual performance of this manufactured eyeball are comparable to those of the human eye.

1.6.2 Microlens With a Dual-Facet Design

Clear eyesight is a crucial adaption in nature that allows living creatures to detect and image their surroundings quickly and accurately. Spherical aberration is a critical technological issue that causes significant picture blurring in lens manufacture. Massive efforts have been undertaken to create microlenses capable of capturing crisp images, including increased manufacturing precision. Nature, on the other hand, has some ingenious answers. As for example, brittlestars have a virtually flawless microlens array made up of a lens structure, which has double faced that is devoid of aberrations and birefringence. To avoid spherical aberration, squid eyes have graded refractive index lenses. In the calcitic skeleton of light-sensitive brittlestars like *Ophiocoma wendtii* the top surface of the microlens is shown in Figure 1.11.

Because of the design of lens which is double-facet, the microlens is entirely corrected for the spherical aberration, and birefringence is avoided by aligning the component calcite's optical axis with the lens axis [87]. Pores around the lens are utilized to carry pigment that controls the amount of light reaching the lens. Modifying the wave beam vectors and their polarization may change connectivity, and the pattern's symmetry, while the laser intensity and exposure time control the pore size and porosity [88]. Taking inspiration from the remarkable optical qualities of the brittle star's microlens and the unique lens shape, researchers have successfully built a biomimetic counterpart using three-beam interference lithography. Hydrogel microlens arrays have also been made using the three-beam interference lithography process by substituting photoresist with cross-linkable-photoacid hydrogel precursors [30]. Soft hydrogels replace rigid microlenses, allowing for continual changes in lens shape and refractive index in response to external stimuli, resulting in intelligent, multipurpose, and adaptable optics [89]. Via template-directed amorphous phase deposition and crystallization [90] the bioinspired method of growing huge micropatterned single crystals offers a promising path to creating artificial brittle star microlens.

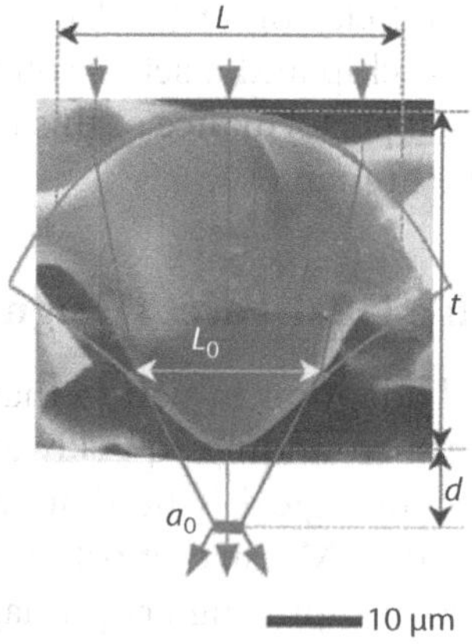

Figure 1.11 High-magnification scanning electron microscope of cross-section of lens L_0. The blue arrow shown in the picture represents the path of the light. The brown lines represents computed lens adjusted for spherical aberration (L is the diameter of the lens, t is thickness, L_0 is the functioning diameter of calcitic lens, d is the location of the focal plane, a_0 is the dimention of the spot) [30].

1.6.3 Fiber Optics in Nature

Optical fibers enable today's data transmission and communication systems. For millennia, fiber optics, a wonderful telecommunications marvel, has florished in nature with sophisticated characteristics. Optical fibers are made up of a core having a very large index value, which is surrounded by a lower-index cladding. As a result, light penetrating the center can completely be reflected within and also light broadcasting over a larger distance with minimum wastage. This is the same pattern seen in glass sponge bio-optical fibers, which transmit light acquired by bioluminescent bacteria to illuminate the whole skeleton in a slightly different way. The primordial creature Euplectella is known for its superior flexibility and mechanical stability while using the least amount of material possible. Deep-sea glass sponges (also known as "Venus flower baskets" because they house a breeding pair of shrimps) live thousands of feet beneath sea level and feature a complex framework made up of glass spicules which are approximately 40 to 70 m thick [91]. They include a lens at the bottom, which can aid the skeleton's bonding with the base. In the spicules, there are three distinct regions: a 2-m-thick core enclosing an organic filament with a refractive index of 1.46; a low refractive index central cylinder covering the core region; and finally, an outer ridged shell with numerous nano-range layers of changing refractive index that expand in thickness from the innermost layer to the outermost layer. Unlike synthetic fibers, these nano-scale layers are linked together using a natural protein, giving them massive flexibility and mechanical strength, helping them to be bent without any cracks.

The skeleton's spicules transmit light differently than unbound fibers at the skeleton's base. This allows the majority of light to flow through the cladding. Another benefit of glass sponge fiber is that it can be made at low temperatures. As a result, the refractive index of the core rises to a greater extent, and sodium ions are easily absorbed into them. Despite synthetic fibers generated at very high temperature, this technique improves their optical characteristics without any birefringence [91]. The V parameter indicated that the fiber's wave-guiding capacity was found to be more than 2.405, signifying multi-purpose behavior. It is also worth mentioning that among all the modes, the lowest-order modes of birefringent multimoded fibers can be quite different from the well-known patterns of multimoded isotropic fibers [51]. Generally, the modal fields are not evenly polarized or circularly symmetric because they are periodically guided circularly symmetric or we can say evenly polarized isotropic fibers. As a result, multimoded anisotropic fibers can exhibit large polarization impacts, which could be useful in practical applications. Fly eyes inspired these fiber creations [92]. When commercial optical fibers were initially conceived and fabricated in the 1950s, they were not bioinspired. Mechanical stability and material elasticity at the nanometer range are two outstanding qualities of naturally occurring optical fibers that should be copied and included into synthetic fibers.

1.6.4 Bioinspired Optical Device

1.6.4.1 Tunable Lenses Inspired by Nature

When it comes to the lens material and the tuning mechanism, tunable lenses have gotten a lot of attention recently. Lenses that employ liquid as their medium can be affected by environmental factors like as gravity and temperature and it also depends on the vibration. This polymer lens (tunable lens) was bonded to a glass substrate and had a fluid chamber and injection port. In the same way as whales can stream liquid in and out of their eye chambers to concentrate light on the surface of retina, an adaptable fluidic lens was developed [93]. The lens' focal length varied between in the range (41–172 mm), with a maximum resolution capacity of approximately 25.39-line pairs per millimetre. Based on the human eye's optical system and how the human eye shifts its focusing point to different distances, Liang *et al.* developed a multilayered tunable lens with a greater focusing range between 14.8 and 30 milimeters and lower lens deviations [94]. The elastic polymer's front surface curvature increased when pressure was applied, simulating the accommodative action of the human lens. In this case, polymethyl methacrylate rings supported the entire device. Cell phone cameras, endoscopes, and machine vision equipment all make use of these small variable focus imaging systems. Many more tunable lenses are being developed that not only mimic the human visual systems but can also even compensate for picture quality via lens [95].

1.6.4.2 X-Ray Telescope

X-rays, unlike visible light, are not refracted as they transit through a substance and are only reflected off a surface when encountered at a grazing angle. Research into astronomical objects that discharge X-ray radiation, part of the electromagnetic spectrum, is required to comprehend the huge-energy cosmos. Roger Angel studied the eye structure of lobsters, working via dazzling angle reflectance, in order to make X-ray imaging conceivable in space [96]. Lobsters, a macruran crustacean, have a superpositional eye with reflective capability and also contain a hemisphere surrounded by outwardly arranged tubes (mainly square-shaped) with reflecting interiors and a field of view of more than 180 degrees, which allows them to perceive objects around them without rotating their heads. Angel created a square cell reflector X-ray telescope with outstanding sensitivity even at very high energies and a vast range of view, akin to that of lobster's eyes. Most importantly, both the reflected and incident beams fall in a common plane. As a result, the angles produced by those incidents and reflected beams are the same. As they go through the real focus, the reflecting rays form two images linearly, as one image is parallel to cell walls while the other image is orthogonal to each other, resulting in a crosslinking pattern. The lenses for X-ray imaging were constructed 100 times longer than their breadth, allowing the system to be very sensitive over a wide energy range. Although the sensitivity is significantly diminished by superimposing linear focus and true focus pictures, the more preferable method is to use moderately bright sources for spectroscopy. Nowadays, X-ray telescopes that mimic the superpositional reflectable eyes of lobsters are being built to a greater extent in order to be launched into space orbit [97].

1.6.4.3 Bioinspired Sensors

Investigating an organism's visual system can lead to new ideas for artificial detectors and imaging systems that outperform current technologies. An ordinary camera and the human eye are both comparable in the capability of image production on the sensor device. Influenced by the optical systems of humans, smart digital camera devices were created with a complementary metal oxide semiconductor (CMOS) or a charge-coupled device (CCD) arrangement under a single aperture lens system [98]. Another focus of research in this discipline is artificial photosensors, which can assist humans to realize their experiences before the natural photosensors have been destroyed. Nowadays, blindness is the most debilitating inability. From the data analysis, the world health organization concluded that around 285 million people have greatly reduced vision and another 39 million people are totally blind all over the world. Artificial sensors and bionic eyes have been developed as a result of research into how the retina of the human eye functions. Actually, the bionic devices activate the unaffected retinal layers electrically through the transplantation of electrodes at the subretinal, supra-choroidal, or epi-retinal level. Second Sight (Sylmar, California, USA) developed the Argus II system, which is the first FDA-approved bionic eye. It is used to treat people who are deprived of eyesight owing to major eye illnesses which damages photoreceptors [99]. As seen below, a tiny video camera is placed on the spectacles together with a video generating device and a retinal prosthesis inserted epiretinally.

In this device, the video generating unit translates camera footage into electronic data for transmission to the retinal prosthesis, which is made up of 60 electrode grids, which can convert information into electrical waves. These waves are then transferred to well neurons in the core retina, and finally, this retina transfers the data to the brain. The Argus II implant enabled users to discern between dark and light in their environment, allowing them to locate and track moving things. A range of bionic technologies, including cortical microstimulation approaches, are being developed to aid in the restoration of vision in individuals with severe visual impairment [100, 101].

1.7 Conclusion

Biomimicry is simply plagiarizing biodiversified nature's tactics and using them to create technology. By putting engineers, biomedical researchers, and biologists together on a single stage, this field of inquiry blurs the boundaries between scientific disciplines, resulting in an interdisciplinary study. Prof. Whitesides correctly pointed out that bioinspiration is a nearly endless supply of ideas with "something for everyone" [102]. Despite the fact that over 14,000 papers have been published since 2010, it is yet believed a young field with plenty of room for new discoveries that will almost certainly lead to the design of materials with interesting functions and applications. Nature has fabricated various structural designs for a vast range of visual phenomena. This may be used to inspire the creation of innovative photosensitive (i.e., optical) substances with a wide range of uses in the future, spanning from everyday life to sophisticated visual technology. New complex structural schemes with exciting optical phenomena will gradually emerge, offering useful models for future optical materials and attracting intense research attention. Due to its considerable potential in a variety of applications, such as consumer electronics, microscopy, medical equipment, and industrial machine vision systems, optical systems that imitate eye functions have caught the interest of researchers. Small sizes, ease of implementation, great adaptability to change environments, and a wide range of functional features are all advantages of eye mimetic optics, which are not accessible with standard optical technologies. However, the absence of well-defined design criteria, as well as the incompatibility of biomaterials and procedures with conventional manufacturing techniques, restricts their use. Through recent improvements in 3D printing and other additive manufacturing technologies, biobased inks may be utilized to imitate biological structures and include biological capabilities and also with the advancement of micro manufacturing technology, using natural inspirations will become more precise, feasible, and versatile. Biomimetic optical material research is substantially accelerated by advanced manufacturing processes such as lithography [30]. These achievements move us closer to realising the potential of materials which are inspired from biology and architectures in day to day life applications. From self-healing structures to artificial brains, bio-inspired research has the potential to change current technology, resulting in a smarter, stronger, and more environmentally friendly future.

References

1. Petsch, S., Schuhladen, S., Dreesen, L., Zappe, H., The engineering eyeball, a tunable imaging system using soft-matter micro-optics. *Light Sci. Appl.*, 5, e16068, 2016.
2. Carpi, F., Frediani, G., Turco, S., De Rossi, D., Bioinspired tunable lens with muscle-like electroactive elastomers. *Adv. Funct. Mater.*, 21, 4152–4158, 2011.
3. Charmet, J., Barton, R., Oyen, M., Tunable bioinspired lens. *Bioinspir. Biomim.*, 10, 046004, 2015.
4. Liu, H., Wang, L., Jiang, W., Li, R., Yin, L., Shia, Y., Chen, B., Bio-inspired eyes with eyeball-shaped lenses actuated by electro-hydrodynamic forces. *RSC Adv.*, 6, 23653–23657, 2016.
5. Jeong, K.-H., Kim, J., Lee, L.P., Biologically inspired artificial compound eyes. *Science, 312*, 557–561, 2006.
6. Durappe, J.W. and Wippermann, F.C., Micro-optical artificial compound eyes. *Bioinspir. Biomim.*, 1, R1–R16, 2006.
7. Floreano, D. *et al.*, Miniature curved artificial compound eyes. *Proc. Natl. Acad. Sci.*, 110, 9267–9272, 2013.
8. Land, M.F. and Fernald, R.D., The evolution of eyes. *Annu. Rev. Neurosci.*, 15, 1–29, 1992.
9. Nguyen, N.T., Micro-optofluidic lenses: A review. *Biomicrofluidics*, 4, 031501, 2002.
10. Mishra, K., van den Ende, D., Mugele, F., Recent developments in optofluidics lens technology. *Micromachines*, 7, 102, 2016.

11. Smith, J.W. and Sheats, J.R., *Microlithography - science and technology*, p. 864, CRC Press, Boca Raton, Florida, United States, 2018.
12. Bernard, G.D. and Miler, W.H., Evidence for visual function of corneal interference filters. *J. Insect Physiol.*, 17, 2287–2300, 1971.
13. Biro, L.P. and Vigneron, J.P., *Photonic nanoarchitectures in butterflies and beetles: Valuable sources for bioinspiration*, pp. 27–51, Wiley, Hoboken, New Jersey, U.S., 2011.
14. Zhang, D.Y., Justis, N., Lo, Y.H., Fluidic adaptive zoom lens with high zoom ratio and widely tunable field of view. *Opt. Commun.*, 249, 175–182, 2005.
15. Werber, A. and Zappe, H., Tunable microfluidic microlenses. *Appl. Opt.*, 44, 3238–3245, 2005.
16. Tadepalli, S., Sloick, J.M., Gupta, M.K., Naik, R.R., Singamaneni, S., Bio-optics and bio-inspired optical materials. *Chem. Rev.*, 117, 12705–12763, 2017.
17. Vukusic, P., An introduction to bio-inspired design. *Contact Lens Spectr.*, 25, 6–13, 2010.
18. Ball, P., Engineering shark skin and other solutions. *Nature*, 400, 507–509, 1999.
19. Geim., A.K., Dubonos, S.V., Grigorieva, I.V., Novoselov, K.S., Zhukov, A.A., Shapoval, S.Y., Microfabricated adhesive mimicking gecko foot-hair. *Nat. Mater.*, 2, 461–463, 2003.
20. Vukusic, P. and Sambles, J.R., Photonic structures in biology. *Nature*, 424, 852–855, 2003.
21. Kinoshita, S. and Yoshioka, S., Structural colors in nature: The role of regularity and irregularity in structure. *ChemPhysChem*, 6, 1442–1459, 2005.
22. Parker, A.R., 515 million years of structural color. *Pure. Appl. Opt.*, 2, R15–28, 2000.
23. Noyes, J.A., Vukusic, P., Hooper, J.R., Experimental method for reliably establishing the refractive index of buprestid beetle exocuticle. *Opt. Express*, 15, 4351–4358, 2007.
24. Land, M.F., The physics and biology of animal reflectors. *Prog. Biophys. Mol. Biol.*, 24, 75–106, 1972.
25. Vukusic, P. and Hooper, I., Directionally controlled fluorescence emission in butterflies. *Science*, 310, 1151, 2005.
26. Bhushan, B., Biomimetics: Lessons from nature - an overview. *Philos. Trans. R. Soc.*, 367, 1445–1486, 2009.
27. Lepora, N.F., Verschure, P., Prescott, T.J., The state of the art in biomimetics. *Bioinspir. Biomim.*, 8, 013001, 2013.
28. Gebeshuber, I.C., Gruber, P., Drack, M., An attempt to reveal synergies between biology and mechanical engineering. *J. Mech. Eng. Sci.*, 223, 2899–2918, 2009.
29. Gebeshuber, I.C. and Drack, M., An attempt to reveal synergies between biology and mechanical engineering. *J. Mech. Eng. Sci.*, 222, 1281–1287, 2008.

30. Yu, K., Fan, T., Lou, S., Zhang, D., Biomimetic optical materials: Integration of nature's design for manipulation of light. *Prog. Mater. Sci.*, 58, 825–873, 2013.
31. Mcphedran, R.C., Nicorovici, N.A., McKenzie, D.R., Rouse, G.W., Botten, L.C., Welch, V. *et al.*, Structural colors through photonic crystals. *Phys. B. Condens. Matter.*, 338, 182–185, 2003.
32. Vukusic, P. and Sambles, R., Shedding light on butterfly wings. *Proc. SPIE*, 4438, 85–95, 2001.
33. Gorb, S.N., *Biological and biologically inspired attachment systems*, pp. 1525–1551, Springer, Berlin/Heidelberg, Germany, 2010.
34. Kolle, M., *Photonic structure inspired by nature*, pp. 121–123, Springer, Berlin/Heidelberg, Germany, 2011.
35. Yoshioka, S., Kinoshita, S., Nakamura, E., Structural color of rock dove's neck feather. *J. Phys. Soc. Jpn.*, 77, 124801, 2008.
36. Vukusic, P. and Stavenga, D.G., Physical methods for investigating structural colors in biological systems. *J. R. Soc. Interface*, 6, S133–S148, 2009.
37. Galusha, J.W., Richey, R., Gardener, J.S., Cha, J.N., Bartl, M.H., Discovery of a diamond-based photonic crystal structure in beetle scales. *Phys. Rev. E*, 77, 050904, 2008.
38. Saranathan, V., Osuji, C.O., Mochrie, S.G.J., Noh, H., Narayanan, S., Sandy, A. *et al.*, Structure, function, and self-assembly of single network gyroid (14132) photonic crystals in butterfly wing scales. *Proc. Natl. Acad. Sci.*, 107, 11676–11681, 2010.
39. Vukusic, P., Sambles, J.R., Lawrence, C.R., Wootton, R.J., Quantified interference and diffraction in single Morpho butterfly scales. *Proc. R. Soc. Lond.*, 266, 1403–1411, 1999.
40. Stavenga, D.G., Giraldo, M.A., Hoenders, B.J., Reflectance and transmittance of light scattering scales stacked on the wings of pierid butterflies. *Opt. Express*, 14, 4880–4890, 2006.
41. Kinoshita, S., Yoshioka, S., Miyazaki, J., Physics of structural colors. *Rep. Prog. Phys.*, 71, 076401, 2008.
42. Lee, R.D. and Smith, G.S., Detailed electromagnetic simulation for the structural color of butterfly wings. *Appl. Opt.*, 48, 4177–4190, 2009.
43. Reid, D.R. and Smith, G.S., A full electromagnetic analysis of the structural color of insects. *IEEE Trans. Antennas Propag.*, 55, 2138–2146, 2007.
44. Bartl, M.H., Butterfly-inspired photonics reverse diffraction color sequence. *Proc. Natl. Acad. Sci.*, 111, 15602–15603, 2014.
45. Radtke, D., Duparré, J., Zeitner, U.D., Tünnermann, A., Laser lithographic fabrication and characterization of a spherical compound eye. *Opt. Express*, 15, 3067–3077, 2007.
46. Kim, J.J., Lee, Y., Kim, H.G., Choi, K.J., Kweon, H.S., Park, S., Jeong, K.H., Biologically inspired led lens from cuticular nanostructures of firefly lantern. *Proc. Natl. Acad. Sci.*, 109, 18674–18678, 2012.

47. Cook, G., Timms, P.L., Spickermann, C.G., Exact replication of biological structures by chemical vapor deposition of silica. *Angew. Chem. Int. Ed.*, 42, 557–559, 2003.
48. Fan, T.X., Chow, S.K., Zhang, D., Biomorphic mineralization: From biology to materials. *Prog. Mater. Sci.*, 54, 542–659, 2009.
49. Chou, S.Y., Krauss, P.R., Renstrom, P.J., Nanoimprint lithography. *J. Vac. Sci. Technol. B: Nanotechnol. Microelectron.*, 14, 4129–4133, 1996.
50. Yang, Y., Song, X., Xiangjia, L., Chen, Z., Zhou, C., Zhou, Q., Chen, Y., Recent progress in biomimetic additive manufacturing technology: From materials to functional structures. *Adv. Mater.*, 30, 1706539, 2018.
51. Lakshminarayanan, V. and Parthasarathy, M.K., Biomimetic optics: Visual systems. *J. Mod. Opt.*, 1–26, 2016.
52. Nilsson, D.E., Vision optics and evolution. *J. Biosci.*, 39, 298–307, 1989.
53. Ivanova, N., Biomimetic optics: Liquid-based optical elements imitating the eye functionality. *Philos. Trans. R. Soc.*, 378, 20190442, 2020.
54. Ramamurthy, M. and Lakshminarayanan, V., *Human vision and perception*, pp. 1–23, Springer, Boca Raton, Florida, United States, 2015.
55. Lee, L.P. and Szema, R., Inspirations from biological optics for advanced photonic systems. *Science*, 310, 1148–1150, 2005.
56. Spering, M. and Montagnini, A., Do we track what we see? Common versus independent processing for motion perception and smooth pursuit eye movements: A review. *Vision Res.*, 51, 836–852, 2011.
57. Völkel, R., Eisner, M., Weible, K.J., Miniaturized imaging systems. *Microelectron. Eng.*, 67, 461–472, 2003.
58. Land, M.F., The optical structure of animal eyes. *Curr. Biol.*, 15, R319–R323, 2005.
59. Qu, P., Chen, F., Liu, H., Yang, Q., Lu, J., Si, J., Wang, Y., Hou, X., A simple route to fabricate artificial compound eye structure. *Opt. Express*, 20, 5775–5782, 2012.
60. Fallah, H.R. and Karimzadeh, A., MTF of compound eye. *Opt. Express*, 18, 12304–12310, 2010.
61. Greiner, B., Ribi, W.A., Wcislo, W.T., Warrant, E.J., Neural organisation in the first optic ganglion of the nocturnal bee Megalopta genalis. *Cell Tissue Res.*, 318, 429–437, 2004.
62. Frederiksen, R. and Warrant, E.J., Visual sensitivity in the crepuscular owl butterfly Caligo memnon and the diurnal blue morpho Morpho peleides: A clue to explain the evolution of nocturnal apposition eyes? *J. Exp. Biol.*, 211, 844–851, 2008.
63. Land, M.F., The optics of animal eyes. *Contemp. Phys.*, 29, 435–455, 1988.
64. Szema, R. and Lee, L.P., *Biologically inspired optical systems*, pp. 309–326, CRC Press, 2005.
65. Land, M.F., Compoud eyes: Old and new optical mechanisms. *Nature*, 287, 681–686, 1980.

66. Sharpe, L.T., Stockman, A., Fach, C.C., Markstahler, U., Temporal and spatial summation in the human rod visual system. *J. Physiol.*, 463, 325–348, 1993.
67. Phan, H.L., Yi, J., Bae, J., Ko, H., Lee, S., Cho, D., Seo, J., Koo, K., Artificial compound eye systems and their application: A review. *Micromachine*, 12, 847, 2021.
68. Zeil, J., Sexual dimorphism in the visual system of flies: The compound eyes and neural superposition in bibionidae (Diptera). *J. Comp. Physiol.*, 150, 379–393, 1983.
69. Aizenberg, J., Tkachenko, A., Weiner, S., Addadi, L., Hendler, G., Calcitic microlenses as part of the photoreceptor system in brittlestars. *Nature*, 412, 819–822, 2001.
70. Li, L., Connors, M.J., Kolle, M., England, G.T., Speiser, D.I., Xiao, X., Aizenberg, J., Ortiz, C., Multifunctionality of chiton biomineralized armor with an integrated visual system. *Science*, 350, 952–956, 2015.
71. Sowards, L.A., Schmitz, H., Tomlin, D.W., Naik, R.R., Stone, M.O., Characterization of beetle Melanophila acuminata (Coleoptera: Buprestidae) infrared pit organs by high-performance liquid chromatography/mass spectrometry, scanning electron microscope, and fourier transform-infrared spectroscopy. *Ann. Entomol. Soc. Am.*, 94, 686–694, 2001.
72. Ali, M.A., *Photorecption and vision in invertebrates*, pp. 115–335, Springer, Berlin/Heidelberg, Germany, 2013.
73. Kiely, P.M., Smith, G., Carney, L.G., The mean shape of the human cornea. *J. Opt.*, 29, 1027–1040, 1982.
74. Sill, T.J. and Recum, H.A.V., Electrospinning: Applications in drug delivery and tissue engineering. *Biomaterials*, 29, 1989–2006, 2008.
75. Ye, J., Shi, X., Chen, X., Xie, J., Wang, C., Yao, K., Gao, C., Gou, Z., Chitosan-modified, collagen-based biomimetic nanofibrous membranes as selective cell adhering wound dressings in the treatment of chemically burned corneas. *J. Mater. Chem.*, 2, 4226–4236, 2014.
76. Roth, L.S., Lundström, L., Kelber, A., Kröger, R.H., Unsbo, P., The pupils and optical systems of gecko eyes. *J. Vis.*, 9, 27, 2009.
77. Sprague, W.W., Cooper, E.A., Reissier, S., Yellapragada, B., Banks, M.S., The natural statics of blur. *J. Vis.*, 16, 23, 2016.
78. Weale, R.A., Problems of peripheral vision. *Br. J. Ophthalmol.*, 40, 392, 1956.
79. Westheimer, G., Pupil size and visual resolution. *Vision Res.*, 4, 39–45, 1964.
80. Nawa, Y., Ueda, T., Nakatsuka, M., Tsuji, H., Marutani, H., Hara, Y., Uozato, H., Accomodation obtained per 1.0 mm forward movement of a posterior chamber intraocular lens. *J. Cataract Refract. Surg.*, 29, 2069–2072, 2003.
81. Dick, H.B., Accommodative intraocular lenses: Current status. *Curr. Opin. Ophthalmol.*, 16, 8–26, 2005.
82. Pierscionek., B.K. and Chan, D.Y., Refractive index gradient of human lenses. *Optom. Vis. Sci.*, 66, 822–829, 1989.
83. Belcher, A.M., Hansma, P.K., Stucky, G.D., Morse, D.E., First steps in harnessing the potential of biomineralization as a route to new high-performance

composite materials Paper presented at Sympos. *Acta Mater.*, 46, 733–736, 1998.
84. Land, M.F. and Nilsson, D.E., *Animal eyes*, pp. 157–235, OUP Oxford, Oxford, England, 2012.
85. Zhang, Y., Liu, J., Qi, K., Liu, Z., Modelling, design and analysis of a biomimetic eyeball-like robot with accommodation mechanism. *Robotics Biomim.*, 861–866, 2013.
86. Schuhladen, S., Petsch, S., Liebetraut, P., Müller, P., Zappe, H., Miniaturized tunable imaging system inspired by the human eye. *Opt. Lett.*, 38, 3991–3994, 2013.
87. Sweeney, A.M., Marais, D.D., Ban, Y.A., Johnsen, A., Evolution of graded refractive index in squid lenses. *J. R. Soc. Interface*, 4, 685–689, 2007.
88. Aizenberg, J. and Hendler, G., Designing efficient microlens arrays: Lessons from Nature. *J. Mater. Chem.*, 14, 2066–2072, 2004.
89. Yang, S., Chen, G., Megens, M., Ullal, C.K., Han, Y., Rapaport, R. *et al.*, Functional biomimetic microlens arrays with integrated pores. *Adv. Mater.*, 17, 435–438, 2005.
90. Aizenberg, J., Muller, D.A., Grazul, J.L., Hamann, D.R., Direct fabrication of large micropatterned single crystals. *Science*, 299, 1205–1208, 2003.
91. Aizenberg, J., Sundar, V.C., Yablon, A.D., Weaver, J.C., Chen, G., Biological glass fibers: Correlation between optical and structural properties. *Proc. Natl. Acad. Sci.*, 101, 3358–3363, 2004.
92. Snyder, A.W. and Rühl, F., New single-mode single-polarisation optical fibre. *Electron. Lett.*, 19, 185–186, 1983.
93. Zhang, D., Lien, V., Berdichevsky, Y., Choi, J., Lo, Y., Fluidic adaptive lens with high focal length tunability. *Appl. Phys. Lett.*, 82, 3171–3172, 2003.
94. Liang, D., Wang, X.Y., Du, J.W., Bioinspired solid-liquid mixed tunable lens with multi-layered structure. *Opt. Eng.*, 54, 065104, 2015.
95. Förster, E., Stürmer, M., Wallrabe, U., Korvink, J., Brunner, R., Bio-inspired variable imaging system simplified to the essentials: Modelling accommodation and gaze movement. *Opt. Express*, 23, 929–942, 2015.
96. Angel, J.R.P., Lobester eyes as X-ray telescopes. *Astrophys. J.*, 184, 84–85, 1979.
97. Hudec, R., Pina, L., Šimon, V., Švéda, L., Inneman, A., Semencová, V., Skulinová, M., LOBSTER- new space x-ray telescopes. *Nucl. Phys. B*, 166, 229–233, 2007.
98. Dean, B.K., Wright, C.H.G., Barrett, S.F., Biomimetic signal conditioning and light adaption for compound vision sensors. *IEE Sens. J.*, 15, 269–278, 2014.
99. Lewis, P.M., Ackland, H.M., Lowery, A.J., Rosenfeld, J.V., Restoration of vision in blind individuals using bionic devices: A review with a focus on cortical visual prostheses. *Brain Res. J.*, 1595, 51–73, 2015.
100. Ong, J.M. and Cruz, L.D., The bionic eye: A review. *Clin. Experiment. Ophthalmol.*, 40, 6–17, 2012.

101. Naik, R.R. and Singamaneni, S., Introduction: Bioinspired and biomimetic materials. *Chem. Rev.*, 117, 12581–12583, 2017.
102. Whitesides, G.M., Bioinspiration: Something for everyone. *Interface Focus*, 5, 0031, 2015.

2

Mimicry at the Material–Cell Interface

Rajiv Kumar[1]* and Neelam Chhillar[2]

[1]*NIET, National Institute of Medical Science, Jaipur, India*
[2]*Department of Biochemistry, School of Medicine, DY Patil University, Navi Mumbai, India*

Abstract
Molecular mimicry transpires through a driving force that can be initiated by artificial biology, designed, engineered structural and functional topographies of natural cells, organelles, or biomolecules by uniting biological and abiotic building blocks. An artificial or assembled material can mimic the numerous physical and chemical features of the natural cellular environment. Because the designing and engineering of biointerfaces are reliant on the abovementioned properties, therefore, it is necessary to investigate neuronal networks in their local environment, which are recognized as long-range biological interaction platforms and have demonstrated great biological importance during the formation of biointerfaces and materialized cellular interactions distinct as interfaces. Apart from that, nature's cellular environment is extremely complex and dynamic. Thus, a new method and phase must be addressed to have a clear grasp of emulation of these prominent possessions. However, it will be extremely difficult to decode mimicking occurrences as they are occurring. When mimicry participates in the events of the natural extracellular matrix at the same time, other highlighted events, characteristics, and attributes (chemistry, topography, and stiffness) are also crucial. By revealing molecular activities, determining cellular localization, and analyzing pathways involving human proteins, the role of mimicry during infection can be explored. Synaptic vesicle trafficking, mitochondrial, growth factors, endocytosis, neurotransmission, axonal transport, and blood-brain barrier features are all routes in which the SARS-CoV-2 protein interfaces with human proteins throughout infection. Virally encoded mRNAs imitate cellular mRNAs to protect the virus from the host's natural immunological limitations. By blocking-specific

**Corresponding author*: chemistry_rajiv@hotmail.com

Inamuddin, Tariq Altalhi and Ashjan Alrogi (eds.) Biomimicry Materials and Applications, (35–82)

targets, the process of mimicking host mRNA machinery that has been initiated by SAR-Co-2 can be stopped.

***Keywords*:** Molecular mimicry, cellular environment, vesicle trafficking, neurotransmission, endocytosis, axonal transport

2.1 Cell and Material Interfaces

Through mimicry, an organism achieves a physiological similarity or similarities with another organism that favors the ability of the mimic to have a survival advantage over it. Besides that, by uniting biological and abiotic building blocks, molecular mimicry happens via a driving force initiated by artificial features of biology, designs, and engineering of structural and functional natural cells, organelles, or biomolecules [1]. The molecules of a pathogen acquire numerous characteristics and surfaces that are equivalent to the peptides of their host through a previously discussed molecular process [2]. When T cells recruit autoimmune disease and initiate an anti-pathogen immune response while interacting or cross-reacting through self-peptides in good physically shapes cells and tissues, the hypothesis of molecular mimicry autoimmunity is initiated. For example, Rheumatic fever is caused by *Streptococcus* when the Epstein-Barr virus's (EBV) Epstein–Barr nuclear antigen 1 (EBNA-1) protein and myelin basic protein will be present in MS cross-react simultaneously [3]. When foreign peptides trigger physiological processes, instantaneously T cells initiate an immunological response. Self-peptides, on the other hand, do not elicit a response.

Cell mimicry, as well as artificial enzymes, artificial mammalian cells, cell–materials interface, and artificial organelles, have been briefly discussed by emphasizing the key aspects and topographies of different mimic routes (Figure 2.1). The number of extremely auto-reactive T lymphocytes in the thymus has been reduced by central and peripheral tolerance pathways and routes. As a result, these transformations inhibit intermediate autoreactivity of these cells in the periphery that are not overstimulated [4]. For that reason, antigenic mimicry is defined as an initiating event for most autoimmune diseases. Various biological molecules and epitopes have been derived from pathogenic agents (bacteria, parasites, viruses, fungi), and a few harmless nutritional molecules (cow milk casein) can act as primary activating antigens and initiate uveitis.

Antigenic mimicry promotes cross-recognition of T lymphocytes and depends on how peptide recognition of antigens does transmit through

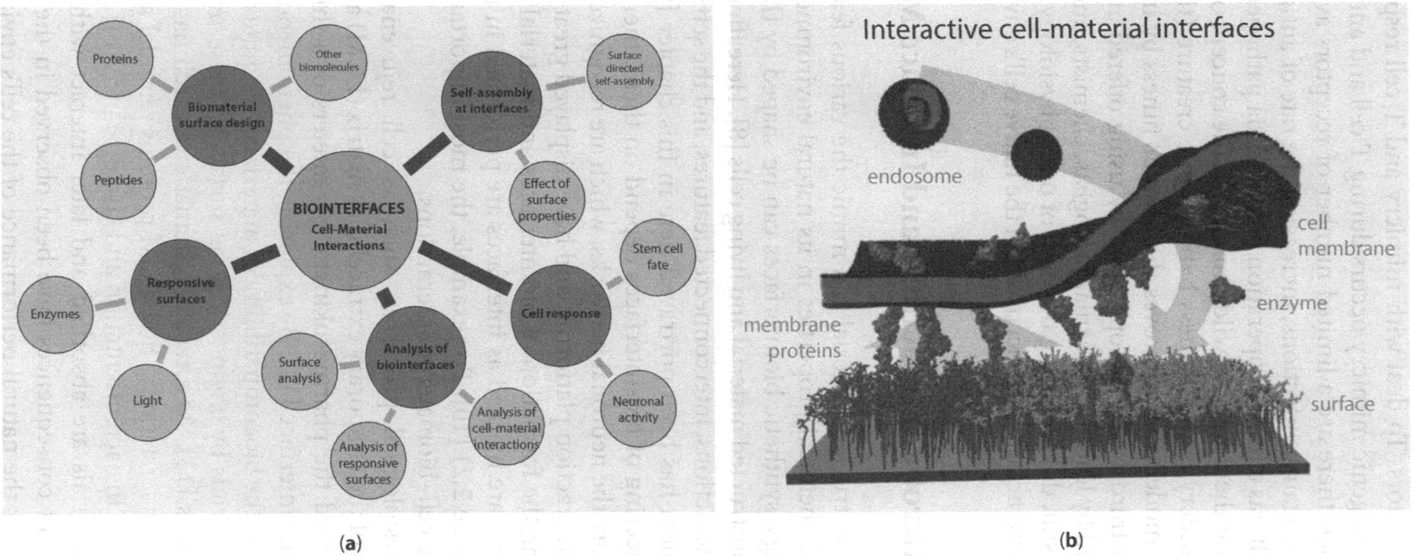

Figure 2.1 (a) Proposed illustration of biointerfaces and classification. (b) Exemplified projected interfaces between cells and materials. Reprinted (adapted) from [5]. Copyright (2022) University of Nottingham.

T lymphocyte receptors. To deal with mimicry and T cell responses, the phenomenon of antigenic mimicry occurs during T cell and antibody recognition [6]. When there are a limited number of receptors available for recruiting defence against infections, increasing the rate of antigen recognition is critical. It has a limited operation patented via pathogen antigens and autoantigens at the time of bewilderment [7]. The concept of mimicry is defined as "Mimicry is the likeness between one creature (the mimic) and another (the model) that improves the mimic's fitness by influencing the behavior of a third party, the operator," by using offered and framed concepts proposed by leading molecular biologists, chemists, and material scientists. As a result, documenting the path of one's physiology requires a grasp of cellular interactions that occurred in the native cell's environment.

2.2 Host-Microbe Interactions and Interface Mimicry

An artificial or constructed material can mimic the various features of a cell's physical and chemical properties in its natural environment. In the same surroundings, synthetic biointerfaces can be shaped by the interaction between a constructed material and living cells [8]. Hereafter, a deeper look at cellular interactions, interconnected features, and the settings of the cellular environment has been portrayed here in this chapter [9]. As the design and engineering of bio-interfaces depend on them, therefore, it is necessary to explore the neuronal networks, which are recognized as long-range biological interaction platforms, and have displayed great biological importance during the formation of biointerfaces and materialized cellular interactions that are marked as interfaces are performed in their local environment (Figure 2.2) [10]. For example, the most important features of these events are cell–biomaterial interactions.

Numerous physiological processes, such as cell rejuvenation, and dynamic creation of the temporary extracellular matrix (ECM) are specific happenings governed the pharmacokinetic parameters of a thin protein layer at the material interface [12]. By exploring the phenomenon of these processes that affect the biocompatibility of materials, it is quite easily possible to design materials by engineering the surface of these materials as per the requirements [13]. As a result, systematic research has been carried out for better understanding how ECM proteins remodel at the cell–biomaterial interface in each orientation and circumstance [14].

Highlighted materials are absorbed and later interact with cell proteins, and significant consequences have been observed in their biological features that alter the natural performance of the cell's environmental

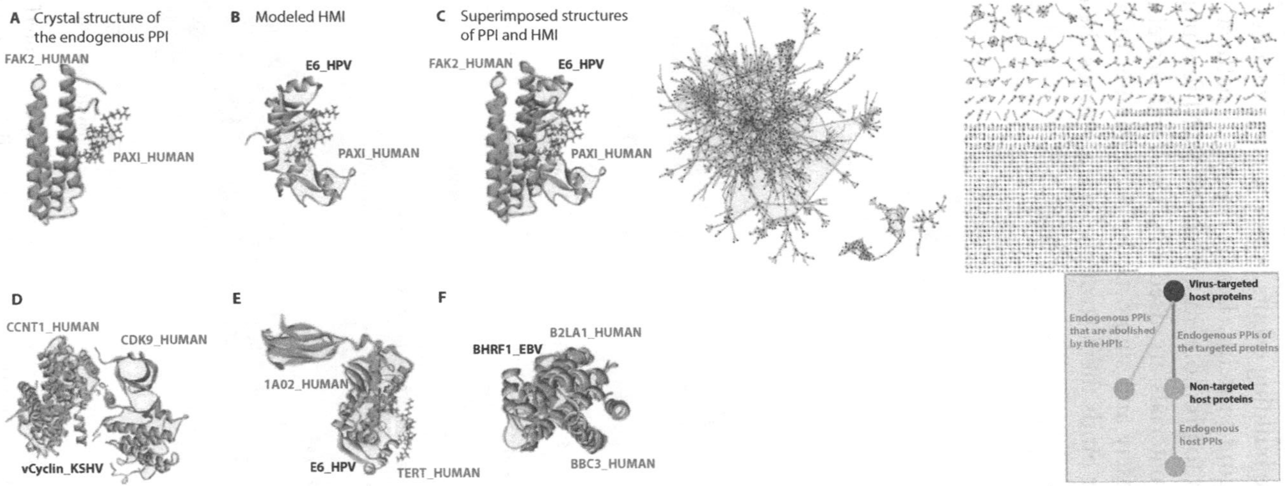

Figure 2.2 Human proteins were represented by cyan and pink proteins, while coviral proteins were represented by grey proteins. Gray proteins network to pink proteins via imitating the cyan protein interface by preventing pink-cyan protein interactions. Reprinted (adapted) with permission from [11]. Copyright (2019) Ruth Nussinov *et al.* Front. Oncol.

surroundings. Temporarily, an interface appeared between cellular proteins and the applied materials, which influenced the assembly of proteins (Figure 2.3). Interface building events promote cell dynamics and involve dynamically rearranging proteins. Lastly, these cell pathways have been labelled as an attempt to promote impermanent ECM functioning. At this point, two types of materials are used that can bind proteins, as one of them binds them loosely and supports the arrangement of a provisional ECM, while the other binds them strongly and promotes the degradation

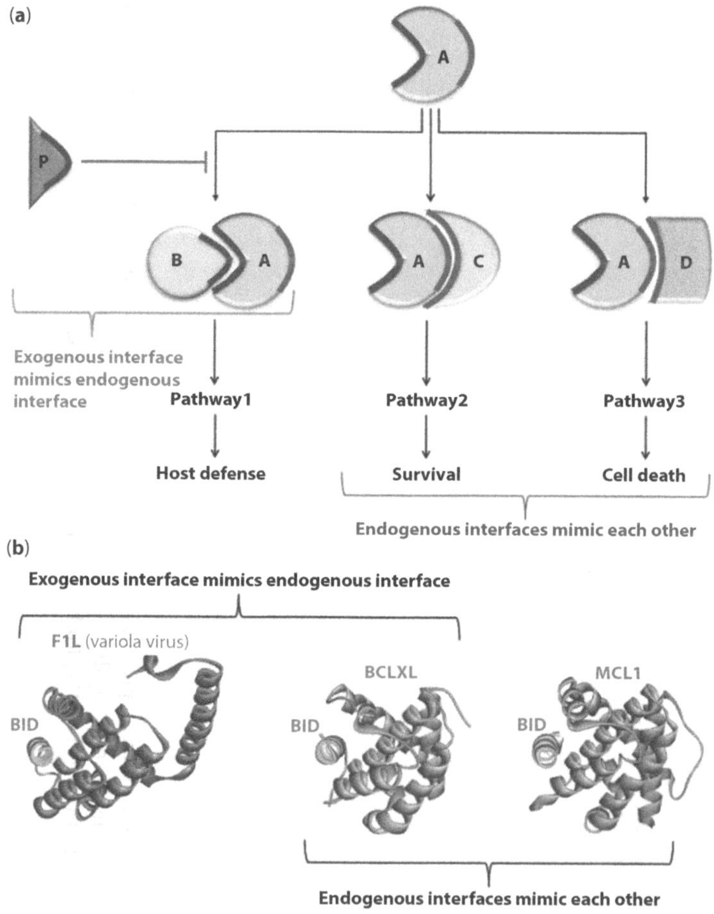

Figure 2.3 Endogenous (intra-species) and exogenous (inter-species) interface mimicry are depicted. (a) The host proteins are marked as A, B, C, and D, while the pathogenic protein is P. (b) Exogenous interface and its mimics phenomenon. Reprinted (adapted) with permission from [14, 15]. Copyright (2017) Ruth Nussinov *et al.* PLOS Computational Biology.

of the ECM [16]. Furthermore, ECM is a dynamic process that takes place in two stages: assembly and degradation.

The biocompatibility of the materials is now linked to the process of reshaping the matrix. Physiological and pathological factors that cause ECM remodelling include normal development, angiogenesis, atherosclerosis, fibrosis, wound healing, ischemia injury, and cancer [17]. Proteolytic remodelling could be used to describe these events. As a result, ECM architecture at the biomaterial interface is regulated by cell transport and physiology, whereas ECM rearrangement is influenced by adsorbed matrix proteins [18]. The underlined process is solely reliant on the correct operational performance of integrin receptors.

2.3 Alterations in Characteristics and Mimicking of Extracellular Matrix

The cell–material interaction transpired naturally to mimic the cellular environment and enhance the chances for a natural interaction of extracellular matrix and materials [19]. During the formation of an interface, the cells promote other activities, including adhering and rearranging adsorbed extracellular matrix proteins on the surface, which are interconnected with the applied material in a pattern similar to fibril. Subsequently, the ECM underwent proteolytic degradation and eliminated excessive ECM for appropriate reshaping (Figure 2.4). At the time of cell–material interface creation, matrix remodelling involves the formation of connections among surface chemistry, matrix rearrangement, cell adhesion, signal degradation, and fibronectin adsorption. At this time, the role of surface chemistry is exposed, and an appropriate style is used in designing new biomaterials [20]. The surface chemistry directly exposes proteolytic routes that come into existence through the cell–material interface formation [21].

These alterations and promotions enhance the features of biomolecule-responsive surfaces that can be used to interact with carbohydrates, enzymes, and peptides. As a whole, reversible bonds are formed with surface-related agents. These chemicals follow non-oval interactions and are modified by stimulation that leads to various changes in chemical surface composition. Nucleotides and carbohydrates serve as reactive materials for modulating a surface's wettability. In it, enzymes change the surface chemistry by irreversibly dividing specific covalent bonds.

Inter-species crossing point mimicry, on the other hand, emerges in proteins from different species and hijacks the similar interface designs [23]. As a consequence, interface resemblance empowers proteins to

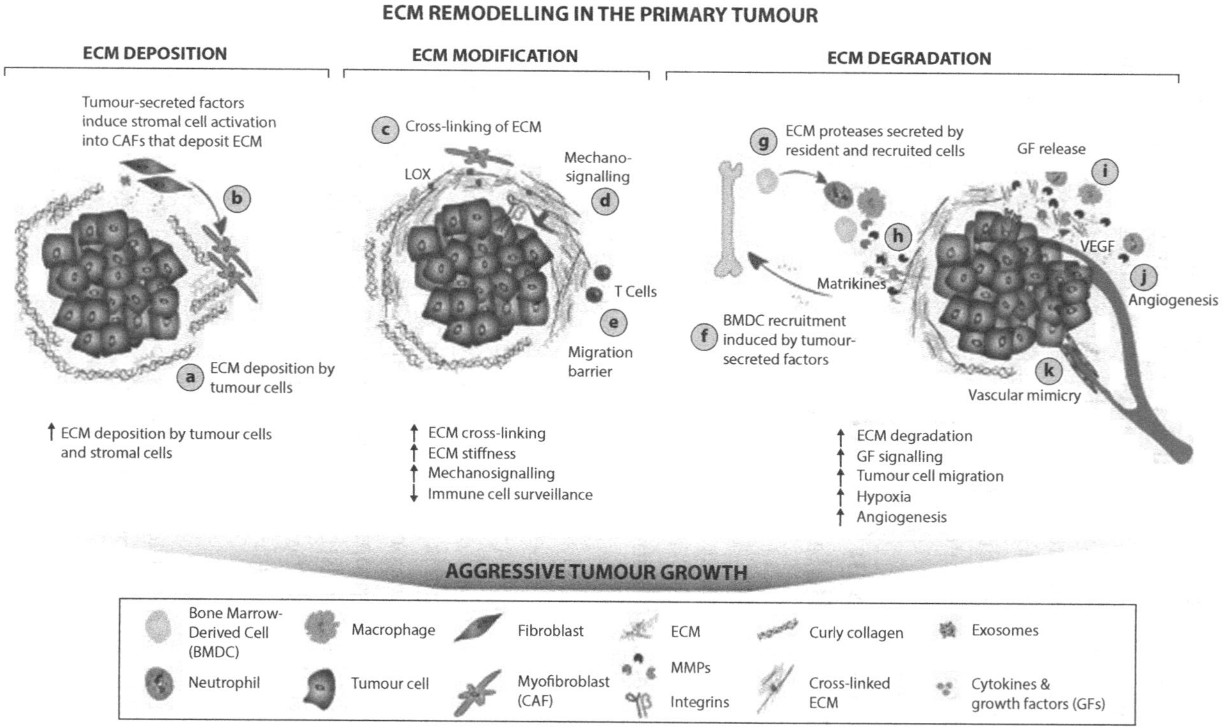

Figure 2.4 An illustration of ECM remodelling and mimicking of extracellular matrix during tumor cell migration. Tumor cells that are stimulated by dense ECM may develop endothelial-like activities and mimic the vasculature that connects blood vessels. Reprinted (adapted) with permission from [21]. Copyright (2020) Juliane Winkler *et al.* Nat Commun.

completely fix a shared target. In addition, an endogenous interface is a linkage between proteins from the same species. Exogenous interfaces are classified as any interface to form proteins from two separate species. By imitating endogenous interfaces, exogenous interfaces make it possible for pathogenic proteins to challenge their host equivalents. Interface mimicry is classified into two types: intra-species and inter-species. When denatured proteins from the same species have a parallel interface, this is known as intra-species interface mimicry [24]. While many bioengineering techniques can be used as a novel tool to initiate matrix degradation in biomedical applications, these events and pathways rehabilitate host signaling networks for improving performance. This occurs, and it has the potential to block or stimulate the host route.

2.4 Mimicry, Manipulations, and Cell Behavior

The chemistry of the ECM present in the cell (protein composition, conformation, and structure) is not static and is constantly changing. The ability of a cell to dynamically adjust biological transformations is important for starting chemical cues at the period of transformation. These events take place at the point when the cell encounters the artificial materials [25]. These changes and reactions are marked as "smart routes" to mimic biological actions [26]. Surface wettability has been used to identify the presence of bio-interfaces, as well as their dynamic surface chemistry, by tempering the accessibility of specific chemical groups on the surface [27]. An outline of the strategies argued in this section to tempt chemical surface alterations in biointerfaces. It is time to have a deeper look at the phenomenon of cell–material interactions and to search for a better way to develop biomaterials that can be applied in advanced regenerative medicine and various applications and are specified in (Figure 2.5). The dynamic elements are the features of mimicry, manipulations, and cell behavior that should be explored [28]. These techniques can be applied to bio-interfaces, where cells react quickly to a surface.

Chemical and physical variables, such as type, dispersion, surface motility, concentration, and adsorbed proteins, are entirely responsible for these characteristics [29, 30]. The protein layer is now affected by changes in the substrate surface topography. Therefore, these features influence the response of cells toward the biomaterials used [31]. The surface chemistry, surface elasticity or stiffness, and surface topography are generally considered the key surface properties that are underlined as key determinants of cell behavior and cell fate. These outlined features and characteristics

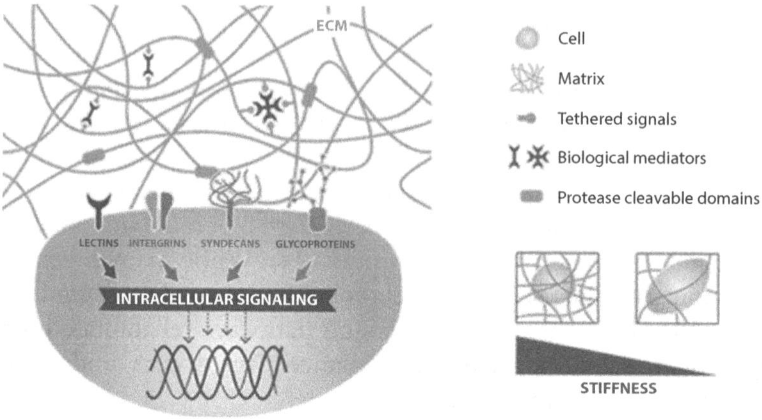

Figure 2.5 ECM cell microenvironment is represented graphically. (a) Interactions between cells and extracellular interactors (biological mediators). (b) Different ECM cell signaling, cell morphology, and shape variables. Reprinted (adapted) with permission from [28]. Copyright (2020) Laura Russo *et al.* Biomacromolecules.

discuss hereby to provide a wide look at mimicry [32]. Overall, the surface properties were explained separately, and have various combinations that are inversely proportional to every biological response.

2.5 Single-Cell Transcriptomics and Involution Mimicry

Tissue engineering is currently being used to replicate the most crucial elements of complicated interface settings and to manage the pathways that regulate cell destiny and functions for appropriate regulation. Advanced tissue engineering, imaging technologies, biosensors, and implants define the essential concepts of innovating biomaterial invention procedures for revealing cell and material interactions [33]. This book chapter exposes diverse key features and beginning opinions to reveal and exploit various routes of cellular interactions that happen when incorporating materials derived via biomedical engineering. Comprehension of the phenomenon of developing cells outside of their natural environments demands a thorough understanding of the biochemical and biophysical topographies that control cell life, as well as interior awareness [34]. The extracellular matrix (ECM), cell organelles, soluble morphogen, growth elements, and a ring milieu that surrounds the cell make up the cell's complex, natural, and dynamic design (Figure 2.6).

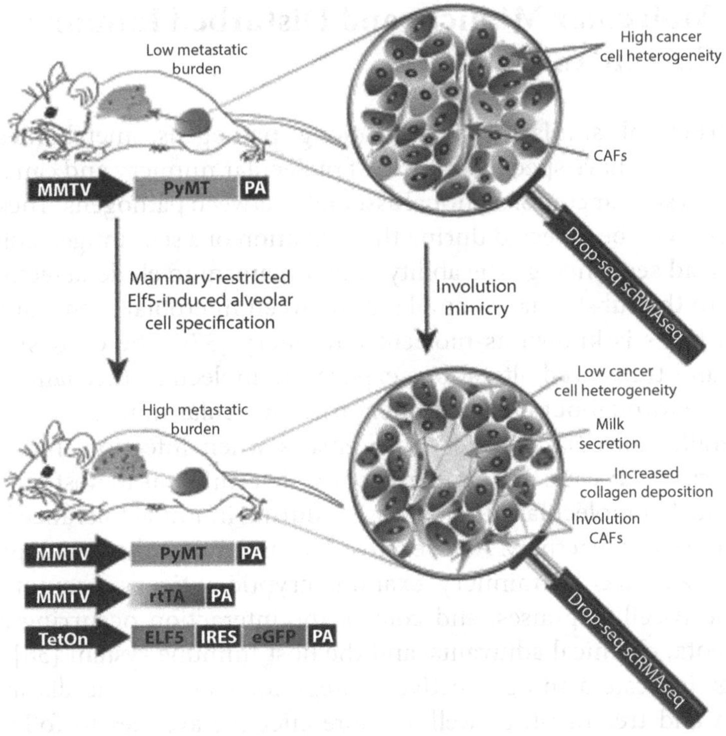

Figure 2.6 Elf5 expression drives involution mimicry, as shown in a molecular model. Reprinted (adapted) with permission from [34]. Copyright (2011) David Gallego-Ortega *et al.* CELL REPORTS, Elsevier.

Cells receive and process information through patio-temporal signaling patterns from the ECM at the same time as remodelling. Beside it, tissue and organ physiology operate through the communication of cells and their microenvironment to pool associations over and done with dynamic and bidirectional interaction. Even so, the complexity existed when the interactions originated between the natural ECM and cells altogether. Hence, a better understanding of the signaling pathways can reveal cell biology [35]. Moreover, exploring these hidden routes is a challenging task, but, these illustrations can help in designing fitting models with a better explanation of interconnected networks. By applying engineered biomaterials, a new approach can be planned to expose the route of deconstruction in cell landscapes [36]. At this point, a strategy should be devised to investigate the impact of separated extracellular matrix constituents on the hosted cells, as well as the cell–material interface and their surface chemistry.

2.6 Molecular Mimicry and Disturbed Immune Surveillance

The structural similarities exist among pathogens, metabolites, and self-structures have specific features of molecular mimicry and can be perceived as a similar conformational assembly between pathogens. These similar events can be detected during the formation of a self-antigen collective amino acid sequencing. The ability of an organism to elude detection and deal with the substantial resemblance between microbial agents and their human hosts is known as molecular mimicry [37]. These physiological routes and transpired alterations expose the molecular mechanisms that take place throughout the phenomenon of molecular mimicry.

Overall, molecular mimicry also ensues when infection initiates the auto-immune disease. The similarity of the determinants to host proteins is a tell-tale that molecular mimicry assists during immune tolerance but not in immune autoreactivity. Autoreactive T-cells, on the other hand, produce secondary molecular mimicry, examine cryptic antigens, promote auto-immune B-cell responses, and control the interaction occurring among microbiota, chemical adjuvants, and the host immune system [38]. These findings indicate a more effective strategy for autoimmune disease prevention and treatment, as well as more effective avenues to follow. The potential repercussions of microbial molecular mimicry are now listed, as well as the causes that particularly target microorganisms by articulating analogous antigens and preventing molecular mimicry [39]. What's more, different pathogens prompt autoimmunity through several stated mechanisms to initiate molecular mimicry. Afterward, specific T cells and pathogens cross-reactivity and initiate autoimmune tissue destruction.

To expose the mechanisms underlying molecular mimicry-induced CNS autoimmunity, different biological routes are advised to have a look at infection pathology that is responsible for initiating autoimmune diseases [40]. One example is molecular mimicry, which occurs when a distant antigen mimics the sequence or assembly of self-antigens. What is more, the initiation of autoimmune diseases does not happen just because of the structural similarities between pathogens and cells that are solely responsible for T cell activation but are also responsible for several other causes (Figure 2.7). One of the postulated mechanisms that were similarly misunderstood was the appearance of twin T cell receptors (TCR) in a single T cell, which was later highlighted as a crucial cause of molecular mimicry [41]. Additionally, the accomplished T cells revealed dual features for both foreign and self-antigens.

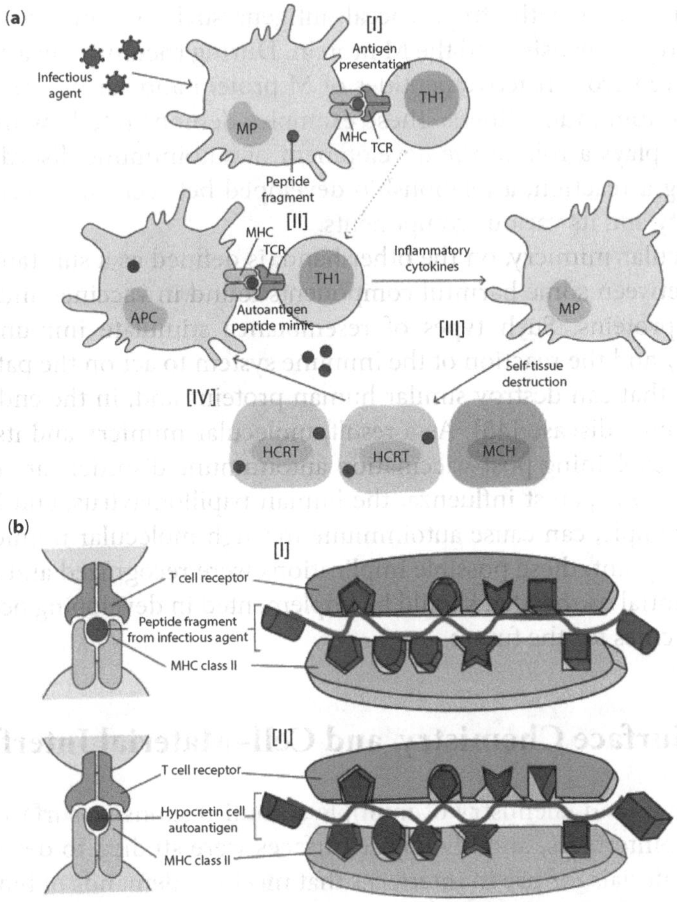

Figure 2.7 (a) The activation of cross-reactive TH1 cells is depicted as a molecular mimicry picture. (b) Molecular mimicry at the synapses between MHC and TCR. Reprinted (adapted) with permission from [42]. Copyright (2014) Fabio García-García *et al.* Sleep Disorders. Hindwai.

As a result, these cells make the host more susceptible to foreign assaults. These pathways, as well as other uncontrollable cell science occurrences, generate an autoimmune response. So, molecular mimicry is discussed deeply to explore to have a persistent understanding of dual TCRs. Molecular mimicry, dual TCRs, and chimeric TCRs stimulate T cell activation, which leads to immunoreactivity and, eventually, autoimmune disorders [43]. Immunologic epitopes have been expressed by molecular mimicry between a microbe and a host, as in a-hemolytic streptococci infection-induced rheumatic fever. These patients' isolated monoclonal antibodies react with streptococcal antigens such as group A carbohydrate antigen, myosin, and the M protein. During rheumatic heart disease, sequestered cross-reactive peptides of M protein and cardiac myosin can trigger autoimmune illness. These examples demonstrate how molecular mimicry plays a role in the development of autoimmune disorders [44]. Initiating a reaction, a relationship developed between the vaccine's susceptibility and its various components.

Molecular mimicry, on the other hand, is defined as a substantial similarity between some harmful components found in vaccines and specific human proteins. Such types of resemblance stimulate immune cross-reactivity and the reaction of the immune system to act on the pathogenic antigens that can destroy similar human proteins and, in the end, induce autoimmune disease [45]. As a result, molecular mimicry and its significance in explaining post-vaccination autoimmune disorders are fascinating. Vaccines against influenza, the human papillomavirus, and hepatitis B, for example, can cause autoimmune through molecular mimicry [46]. As a final point, these possible implications were recognized and featured as a potential source that should be implemented in developing better, and safer vaccines for the future.

2.7 Surface Chemistry, and Cell–Material Interface

The physics and chemistry of multiple stimuli-responsive surfaces, interactive biointerfaces, and 3D object surfaces were studied to develop better biomaterials for use in interfaces that meet the demands of biomedical applications [47]. Although several theories for describing the cell–material interface have been explored, they have yet to be applied to biological claims. Complex 3D materials with responsive surfaces can be applied to build a cell–material interface that closely resembles natural surroundings [48]. In current scenarios, novel biomaterials have been designed to deal with the complex features of the natural cellular environment and mimic cellular features and concerned responses. The discovery of stimulus-responsive materials can contribute to incorporating dynamic and reversible elements into originated biointerfaces (Figure 2.8). Cell–material interaction is a complicated process that happens at a newly developed interface (biointerfaces). However, these biomaterials can be used in the design of different responsive biointerfaces and can be used to explore new opportunities in various applications in biomedical medicine. This chapter demonstrates the need to have perfect-shaped biomaterials that can be imposed on responsive biointerfaces as per the existing requirements.

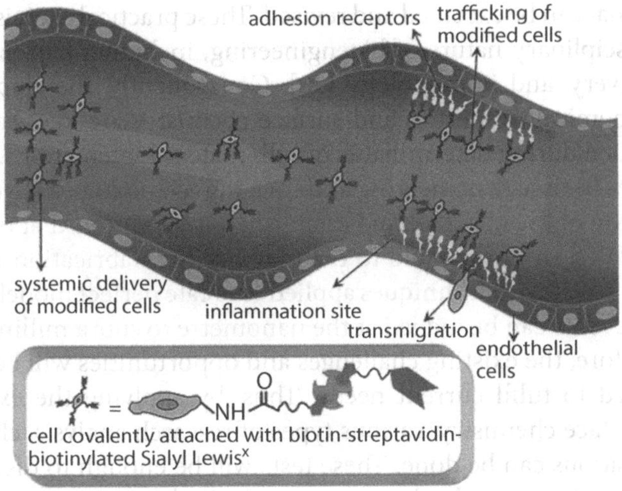

Figure 2.8 Cells from the blood are targeted to inflamed endothelium receptors; cell surfaces are illustrated respectively, employing covalent and noncovalent chemistry. Reprinted (adapted) with permission from [49]. Copyright (2008) Debanjan Sarkar *et al.* Bioconjugate Chemistry, American Chemical Society.

The physics and chemistry of multiple stimuli-responsive surfaces, interactive biointerfaces, and 3D object surfaces were investigated to have better planned and established biomaterials for use in interfaces as per the requirements of biomedical applications significantly [47]. A perfect establishment of an interface between the cells of the patient and the components of the device will demonstrate the worth of the approach. As a result, the proper interaction of the cells with the biomaterial might be considered the most important factor of effective execution. Aside from that, the natural cellular environment is complex and dynamic. Therefore, to have a proper understanding and emulation of these featured possessions, a different route and phase have to be considered. However, while decoding mimicry events is taking place, it will be a huge challenge [50].

The underlined events, features, and properties (chemistry, topography, and stiffness) are important when mimicry partakes in the events of the natural extracellular matrix simultaneously existing. Thus, a proper analysis of the origin of biointerfaces transpires when cell–material interactions are also considered an important event and phenomenon. Based on current practical explanations of these events, cell and material interface-based applications have been investigated [51]. Accordingly, biomaterial creation, surface engineering, and modification of biointerfaces have instigated cell response to surfaces, and the process of self-assembly keeps going at interfaces that can be executed and revised. These practical exercises explain the interdisciplinary nature of bioengineering, including biosensing, controlled delivery, and immunology [20]. Consequently, the elucidation of dynamic chemical properties and surface chemistry are key features that can be applied during determinants of cell–material interactions.

Advances in tissue engineering and imaging technologies have explored the formation of the interface and established interactions that transpire at different scales, from the nano to the macro. Microfabrication and nanofabrication are the two techniques applied to create perfect models in which molecular events can be judged at the nanometre to supra millimetre scale [52]. Therefore, the existing challenges and opportunities will be searched and outlined to fulfil current needs. Thus, by applying the expertise of material surface chemistry, various types of research analysis, elucidation, and explanations can be done. These tests will be enough to discourse the existing biomaterial and healthcare research challenges.

2.8 Cell Biology and Surface Topography

Through reversible chemistry, self-assembled materials change the appearance of surfaces. Supramolecular structural motifs can be created by replicating the noncovalent interactions that proteins and endogenous peptides have. Covalent bonds are modified as a result of chemical surface modifications. As a result, self-assembly produces a characteristically irreversible and reversible display [53]. Zipper molecules are heterodimer complexes that can be created using polypeptides with predefined amino acid sequences. Arg-Gly-Asp (RGD) sequences can sometimes be seen terminating polypeptide zipper molecules and attaching them to a surface [54]. Hence, dimerization with the subsequent zipper arrangement enables noncovalent alterable surface chemical modification (Figure 2.9). Cell adhesiveness of the surface can be re-established competitively by introducing unmodified matching zipper sequences, and these events can subsequently swap the poly(ethylene glycol) (PEG) conjugates.

An RGD-based peptide that investigates host-guest chemistry benefits from a modified self-assembly technique [50]. When a minor molecule is enclosed within a bigger molecule and creates a cage-like molecule, a supramolecular complex is generated. Azobenzene, for example, is a photoresponsive molecule that forms a host-guest connection with -cyclodextrin in its trans-state. When azobenzene is exposed to light, its conformation is occasionally changed to the *cis* state. These stereochemical changes cause conformational changes in the cyclodextrin cavity, making it impossible

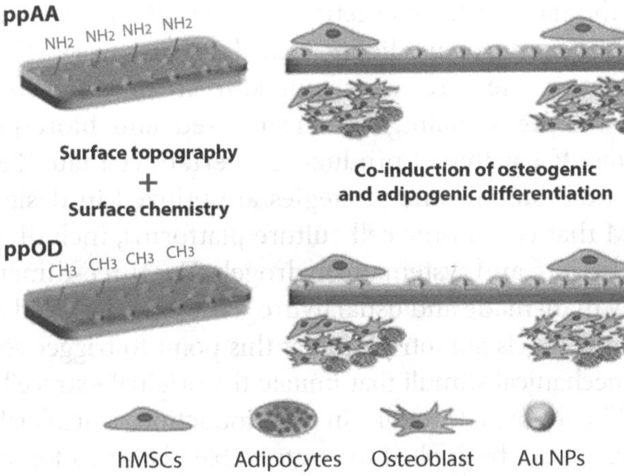

Figure 2.9 An understanding of the individual effects of surface chemistry or topography on cell behavior. Reprinted (adapted) with permission from [55]. Copyright (2021) Kvasilev *et al.* ACS Applied Materials and Interfaces, American Chemical Society.

for azobenzene to enter [56]. By attaching the peptide sequence Gly-Arg-Gly-Asp-Ser (GRGDS) to azobenzene in the presence of light, the chemical features of the cyclodextrin exterior can be manipulated. The RGD sequence identified allows HeLa cells to communicate with azobenzene in its trans-state bound to the surface.

Cell separation can occur when the peptide sequence is detached from the surface, as an irradiation of the peptide at 365 nm, which can break down the supramolecular complex. As a result, surface chemistry can be changed through conformational surface modifications, which can be induced electrical potential to the surface. Chemical functionalities exhibited at the surface can also be altered to achieve these transitions [57]. The attachment of the protein and bacteria to the surface can be adjusted by changing the charge and hydrophilicity one by one or both on the surface. To hinder cell adherence in the absence of an electrical potential, the stimulating molecules stretched and fortified the peptides from cells mounting on the surfaces.

2.9 3D Extracellular Matrix Mimics and Materials Chemistry

The strategies developed for designing new biomaterials are gaining attention, and the designed materials are performing naturally with the cells and tissues at the nanoscale. Synthetic 3D extracellular matrices are used in developing regenerative medicine, and drug discovery. The scaffolds' chemical and physical features were applied to create extracellular matrix mimics based on the signaling chemicals used and bioresponsive cell microenvironments capable of producing a certain cell fate [58]. Overall, various synthetic materials and strategies are utilized in designing functional 3D ECM that can mimic cell culture platforms, including artificial, permissive hydrogels, and systematic hydrogels. For three-dimensional cell culture, equally man-made and usual hydrogels were employed as scaffolds [19]. Synthetic hydrogels are introduced at this point to trigger complex biochemical and mechanical stimuli that imitate the original extracellular matrix (Figure 2.10). Finally, breakthroughs in the production of artificial–biological hydrogel amalgams can be applied to create stable platforms for studying cell physiology.

Complex three-dimensional (3D) tissue analogues are being developed for use in a variety of tissues and organs. To simulate the native ECM structure at the macroscale, a unique electrospinning approach is employed to produce a three-dimensional fiber-hydrogel complex [59]. The phenomenon of mimicking the native extracellular matrix is well considered in supramolecular chemistry and can be applied theoretically and practically as a possible tool in developing renewing medicine and drug discovery.

Nonetheless, the next-generation synthetic materials were offered, which have hierarchically designed supramolecular moieties and can mimic key features of the ECM. These tools and designs are highly dynamic and display both a scene and reactivity for generating external stimuli [24]. Then, again, the development of such desired materials remains a challenging task. Furthermore, the supramolecular complexes were created employing

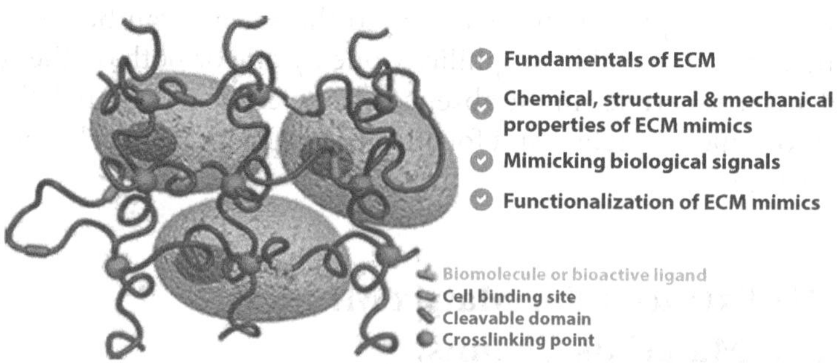

Figure 2.10 Underlying role of 3D extracellular matrix concepts and role of materials chemistry in stem cell fate. Reprinted (adapted) with permission from [28]. Copyright (2020) Laura Russo *et al.* Biomacromolecules, American Chemical Society.

motor amphiphilic materials that can imitate the structural and morphological characteristics of hydrogels as well as ECM properties. The innate dynamic activities of highlighted assemblies and moieties were also revealed, which can be exacerbated by molecular motions at the microscale.

2.10 Microbe Interactions and Interface Mimicry

A deeper analysis is important at the molecular level of host-microbiota interactions and to gain insights into the mechanisms of infections. Furthermore, a proper investigation of the crosstalk that transpires between the host and the microbiota can reveal many hidden features and facts [60]. Furthermore, microbes' metabolic pathways will disclose their involvement in health and illness. Integrated superorganism setups have the specifics of the host and microbiota interactions. Hence, they should be specifically investigated. These findings will uncover the complete patterns of their code [28]. This elucidation will uncover the most fundamental theories and will explain the routes of modulation of immune surveillance (Figure 2.11). The effective microbial effectors can be identified and uniquely by illuminating different cellular frameworks and networks of structural host-microbiota interactions [61]. These interpretations can be utilized to target previously unknown parts and features of various host nodes, as well as disrupting endogenous host interfaces. These theoretical and practical insights will aid in the research and discovery of previously unknown aspects, such as how mutations in the host or microbial proteins affect interaction [62]. Master host cells were also identified as controller nodes and modules that the microbes manipulated to produce anomalous behavior in structural host-microbiota interactions.

Collectively, the elucidation of pathogenic mechanisms can help design accurate therapeutics [64]. But, many studies are persisting in searching for experimental techniques that can explore the existing network at a large scale to get a proper blueprint of host-microbiota interactions. Moreover, by using computational techniques, the biological routes of interactions can be predicted in a better way and can be mapped to the host cellular networks [65]. These findings may open up new avenues for research and provide answers to issues such as "What channels and networks exist in exhibiting and pathogenic microbiota modulation that influences host signals and is generally cross-species?" As signaling pathways control cell behavior, functioning, and transport of the cell's reaction toward its concerned and interconnected cellular environment, in the same way,

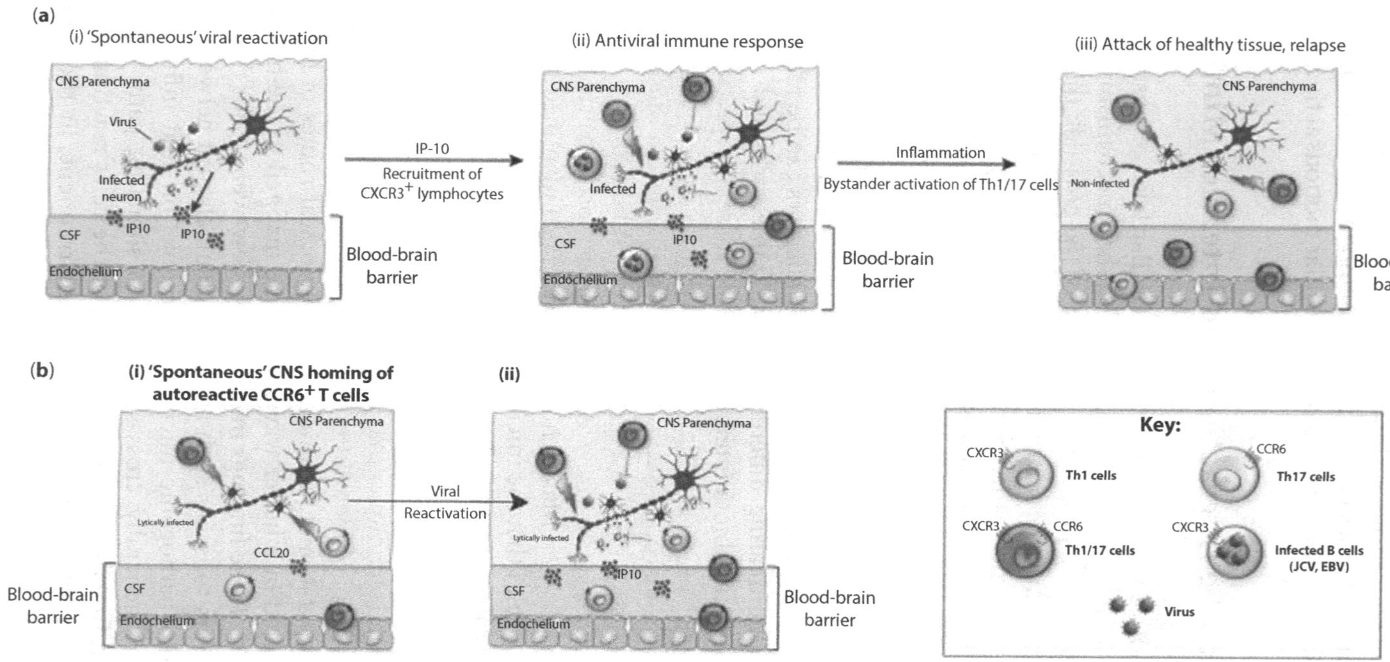

Figure 2.11 (a) Antiviral T lymphocytes examine the central nervous system (CNS) in healthy people. (b) Alternatively, autoreactive cells. Reprinted (adapted) with permission from [63]. Copyright (2017) Jens Geginat *et al.* Trends in Immunology, Science Direct.

different pathogens outwit the originating responses by influencing host signaling.

Microbe interactions undermine cell physiology and hijack host pathways to damage the host defence by imitating the requisite outsides of host-prearranged proteins. Then, different folding mechanisms generally act together via parallel protein-protein interface designs. Additionally, these similarities persisting in the binding surfaces allow the protein of a microbe or pathogen to contend with a host target protein. Here, the host protein hub grips the pathogenic surrogate but does not bind a host-encoded partner [66]. Finally, the repurposing of the host pathways, the immune response, and therefore the shifting of the cell signaling landscape

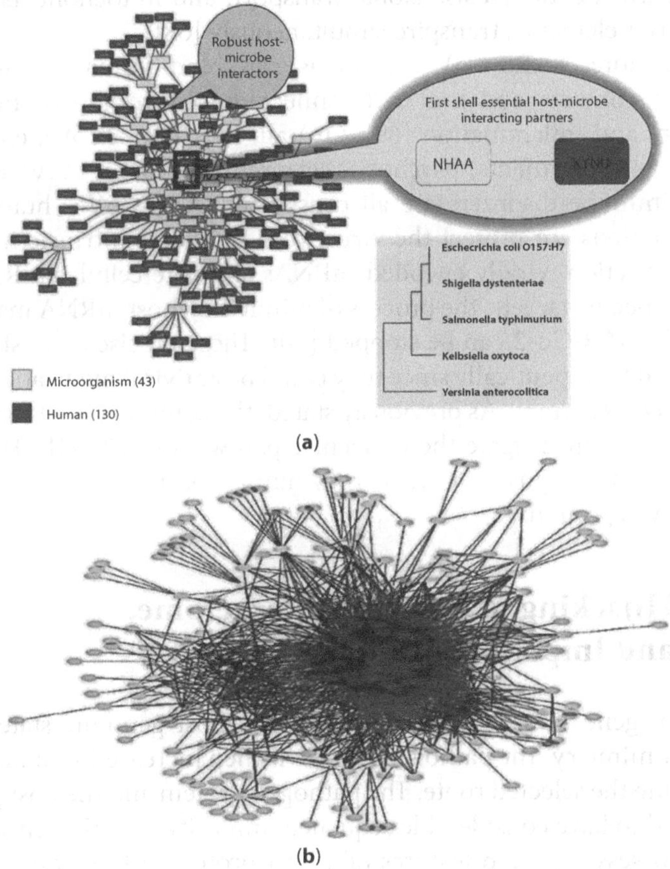

Figure 2.12 (a) Essential host-microbe interacting proteins. (b) Host-microbe disease network. Reprinted (adapted) with permission from [67]. Copyright (2020) Shivani Sharda, scientific reports, nature.

followed each other. These events initiate persistent infections and, later on, can cause cancer by altering the main signaling pathways (Figure 2.12).

To map the renewed host-pathogen and linked interface setup, as well as its basic specifics, a deeper understanding of pathogenic mechanisms, is necessary. The physiognomies of molecular mimicry that happened in pathogen-host shirking, as well as how molecular mimicry processes transpire and materialized during evolution, should be highlighted [62]. Elucidation of molecular functions, detection of cellular localization, and analysis of pathways concerning human proteins can be useful in exploring the role of mimicry during infection. When SARS-CoV-2 induces infection, its protein interacts and communicates with human proteins. Various biological routes, such as synaptic vesicle trafficking, neurotransmission, growth factors, endocytosis, axonal transport, and mitochondrial blood-brain barrier elements, transpire simultaneously [68].

Furthermore, SARS-CoV-2 proteins can bind human proteins via peripheral contacts and are well-connected to thrombosis, metabolic regulation, and inflammation [69]. Overall, Delirium, stroke, encephalitis, sensory impairments, psychosis, seizures, peripheral nerve illnesses, and autoimmune disorders are all caused by SARS-CoV-2-human protein interactions. To protect the virus from the host's intrinsic immunological restriction, virally encoded mRNAs simulate cellular mRNAs. By blocking-specific targets, the process of mimicking host mRNA machinery initiated by SAR-Co-2 can be stopped [70]. These precise areas should be investigated therapeutically since they could be antiviral medication marks for COVID-19 therapy. As previously stated, the computational techniques can be used to investigate the molecular pathways of COVID-19-related neuropsychiatric signs that stem from human protein interactions with SARS-CoV-2 proteins.

2.11 Hijacking of the Host Interactome, and Imperfect Mimicry

Horizontal gene transfer is used to obtain a host genomic state during molecular mimicry. The pathogen's fitness is then increased simultaneously by following the selected route. The pathogen protein and the host proteins were found to have considerable sequence similarity. A pathogen-encoded protein possesses certain features of a host protein at this point, and the mimicry is then completed by unplanned mutations and usual choices. As a result, similar events have structural similarities but minimal sequence homology [71]. What is more, viruses and bacteria apply the phenomenon of molecular mimicry for host recognition when the binding of host receptors transpires. The pathogen mimic and the corresponding host receptor are both involved in indirect interactions due to the imitation of native cellular ligands. The step sought after by bacteria and viruses, ligand mimicry, was observed during silico studies (Figure 2.13).

To obtain an edge or compete for similar resources, an organism, cell, or protein imitates another. In general, infections use molecular mimicry in the form of pathogen-encoded constituents that imitate host elements to co-opt or impair host functions and traits [72]. Pathogens have an edge here. For example, herpes and poxviruses encode host immune managers such as chemokine and cytokine, and all infections achieve immune evasion by making these changes [73]. In the same way, bacterial virulence

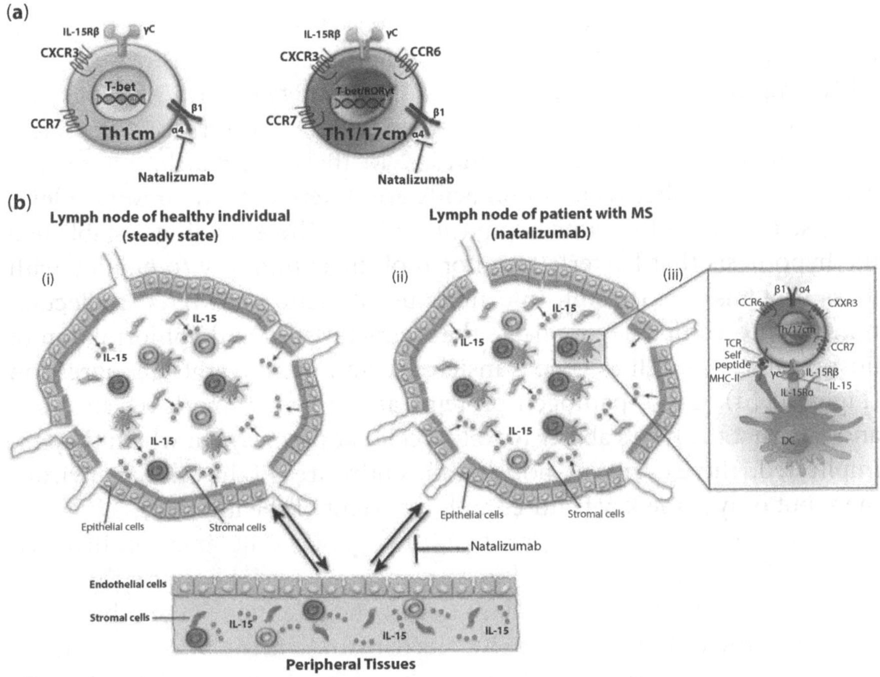

Figure 2.13 A hypothetical method of protective central memory competition. (a) Th1CM and Th1/17CM cells displayed higher levels of the transcription factor T-be. Reprinted (adapted) with permission from [63]. (b) In healthy individuals (i) Th1CM and Th1/17CM cells existed in equilibrium (b) (ii) autoreactive Th1/17CM cells expand via virus-specific Th1CM cells, (b) (iii), Th1CM cells are not properly fit under these conditions and die by neglect. Copyright (2017) Jens Geginat *et al.*, Trends in Immunology, Science Direct.

factors interfere with normal host biological processes that proceed by mimicking vital host proteins.

Molecular mimicry is underlined as the main cause that is responsible for establishing the relationship between autoimmune syndromes and specific infections. For example, multiple sclerosis originates because of Epstein-Barr virus (EBV) infection [75]. The cross-reaction, which takes place between the antibodies and autoimmune syndromes, reduces the impact of the infection caused by the infected agent on host protein [76]. The discovered findings ensure the presence of bacterial effectors and promote the involvement of the newly discovered interface, which imperfectly mimics the eukaryotic cellular environment [77]. The evolutionary pressure exerted by effectors at this point increases the number of networks with host cells while concurrently reducing the numeral exchanges with added pathogen proteins.

By mimicking eukaryotic interfaces, originators can be transformed into stickier, and potentially enhance interactions with other pathogen proteins [78]. Overall, these types of lethal interactions disturb cell viability and enhance infectivity. To distinguish between bacteria-bacteria and bacteria-host interactions, the highlighted effectors exploit defective mimicry of eukaryotic boundaries. In these events, the interaction patterns existing between amino acids are different at the interface level because they have the same composition [79]. These readings established the hypothesis that bacteria monitor molecular mimicry to employ with mass proteins. As a result, infections use a tactic known as "molecular mimicry" to completely replicate the characteristics and conformation of host proteins, as well as gene transference and/or convergent progression (Figure 2.14). These pathogen proteins are driven to mark the host site and obstruct the host's ability to follow cellular pathways in which they are implicated. These events do not alter the entire area of the cellular environment but only some key features and the extent of the interface.

At this point, the pathogen proteins disrupt host cellular events involved in cell adhesion and invasion. These events are crucial because they start the infection process. As a result, these channels define the hypothesis that underpins the tight host selectivity mechanism. During infections caused by these dangerous bacteria, the same theory has been used [80]. A mimicry is a molecular event that occurs on a case-by-case basis and is based on the arrangement or near structural resemblance. Pathogen effectors mimic eukaryotic complexes when imperfect mimicry of protein interfaces comes about to enhance the adaptation in the host. The pathogen mimics the involvement in the alteration of its proteins by the pathogen mimics (Figure 2.15).

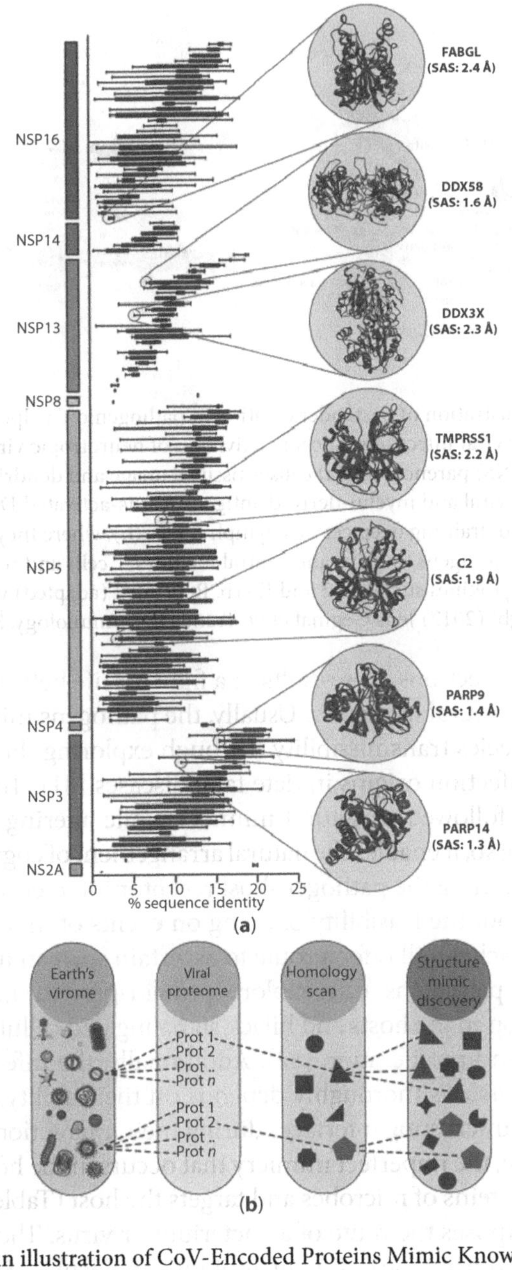

Figure 2.14 (a) An illustration of CoV-Encoded Proteins Mimic Known Regulators of Human Immune Response to Infection. Reprinted (adapted) with permission from [74]. Copyright (2021) Sagi D. Shapira *et al.* Cell Systems, © 2020 Elsevier Inc. (b) Graphical schematic of intermediate, sequence-detected structural templates and the strategy implemented to reveal virus-encoded protein mimics. That are geometrically similar to viral proteins for revealing viral mimicry [74]. Copyright (2021) Sagi D. Shapira *et al.* Cell Systems, © 2020 Elsevier Inc.

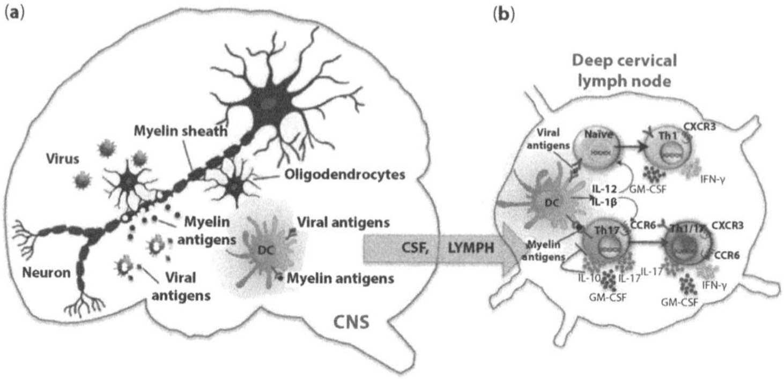

Figure 2.15 An illustration of bystander creation of pathogenic T helper (th)-1/17 cells may be tempted by viral infections. Local reactivation of neurotropic viruses in the central nervous system (CNS) parenchyma (a) causes tissue damage and dendritic cell (DC) absorption of both viral and myelin-derived antigens. Virus-activated DCs subsequently travel via the CSF to draining deep cervical lymph nodes (b), where they offer both viral and myelin-derived antigens to naive and central memory T cells and release antiviral and proinflammatory cytokines such IL-12 and IL-1b. Reprinted (adapted) with permission from [63]. Copyright (2017) Jens Geginat *et al.* Trends in Immunology, Science Direct.

During viral infections, this results in a fight for life between the host and the pathogen at several interfaces. Usually, the pathogens mimic host receptors and interspecies transmissibility. Through exploring these main events, the emerging infection origins initiate fatal diseases [81]. The bacterial and viral pathogens followed structural mimicry while altering the native host ligands. By doing so, it enables the natural arrangement of cognate host receptors [74]. By screening the pathogen-host receptor interactions, a prediction can be made about the feasibility of going on events of molecular mimicry, and these approaches will offer a route to ascertain such an interaction.

Overall, the pathogens can explore protein-protein interactions and inductee infection in the hosts and hijack signaling and cellular pathways for their existence and proliferation [82]. Additionally, the infectious potential of bacteria and viruses thoroughly depends on their ability to adapt to the geometry of the interacting interfaces during their interaction with host proteins. Meanwhile, the imperfect mimicry that occurs in the host cell interface promotes the proteins of microbes and targets the host (Table 2.1) [83]. This consideration exposes the route of a bacterium or virus. These findings can be implemented during the design of antibiotic molecules [84].

Furthermore, SARS-CoV-2 hijacks host cell machinery by imitating epithelial ion channels. It is immunopathogenic in general. The SARS-CoV-2 spike protein (S1/S2) and its cleavage site have a unique protein sequence. Similar events occurred in the epithelial sodium channel subunit (ENaC). The existence of structural mimicry of the spike protein of the SARS-CoV-2

Table 2.1 Different aspects of recent investigation, interface mimicry, single-cell transcriptomics, manipulations, cell behavior, cell biology, and surface topography, host-microbe interactions.

S. no.	Various features of cell–material interfaces	Impact of surface chemistry and surface topography on biomaterial–cell interface	Ref.
1.	Mimicking cell–cell interactions at the biomaterial-cell interface for control of stem cell differentiation	The Notch system regulates tissue development and stem cell destiny by making surface changes that finally imitate the extracellular matrix.	[85]
2.	Cell/material interfaces: influence of surface chemistry and surface topography on cell adhesion	Biological mechanisms behind the cell's reaction to topography, as well as the impact of surface chemistry and energy on cell adhesion	[86]
3.	Cell and material interface: advances in tissue engineering, biosensor, implant, and imaging technologies	Advances in tissue engineering, imaging technologies, biosensor investigates this relationship and how it can be controlled at length scales ranging from nano to macro.	[87]
4.	Chemical communication between synthetic cell/living cell interface	Synthetic biology and bioengineering based on processes of distributed communication, signal processing, nested communication, network-based contact-dependent communication, and physical embodiment for proper chemical communication in consortia of synthetic and living cells.	[88]

(*Continued*)

Table 2.1 Different aspects of recent investigation, interface mimicry, single-cell transcriptomics, manipulations, cell behavior, cell biology, and surface topography, host-microbe interactions. (*Continued*)

S. no.	Various features of cell–material interfaces	Impact of surface chemistry and surface topography on biomaterial–cell interface	Ref.
5.	Responsive cell–material interfaces	Biomaterial interfaces will be included, including numerous stimulus sensitive surfaces, 3D object surfaces, and interactive biointerfaces.	[22]
6.	Molecular mimicry in innate immunity	The transcription factor NF-B, which governs the manufacture of pro-inflammatory cytokines, is activated by adaptor.	[89]
7.	Expansion microscopy and imaging proteins	Cell adhesion, migration, and differentiation have been influenced by surface topography on the scale of tens of nanometers.	[90]
8.	Cell–cell interfaces directs cell function	Interactions between motile immune cells to long-term cell–cell contacts in epithelia, cell–cell interfaces have been observed throughout multicellular animals.	[91]
9.	Control cell–material interfaces	Supramolecular hydrogels, can be used to manipulate cell behavior and function, ultimately influencing cell fate.	[92]
10.	Vascular mimicry and molecular interactions	Signalling molecules dynamic cancer stemness and autophagy. Vascular mimicry increases cancer malignancy and metastasis.	[93]

(*Continued*)

Table 2.1 Different aspects of recent investigation, interface mimicry, single-cell transcriptomics, manipulations, cell behavior, cell biology, and surface topography, host-microbe interactions. (*Continued*)

S. no.	Various features of cell–material interfaces	Impact of surface chemistry and surface topography on biomaterial–cell interface	Ref.
11.	Cell mimicry and nature-inspired entities	Synthetic biology and molecular building pieces are combined to produce proto cells that resemble biological features.	[94]
12.	Mimicry as an interface between psychology and evolution	This "mimicry" can't be explained by evolutionary inertia; it has to be adaptive. (As a side note, the term "mimicry" was coined before Darwin.)	[95]
13.	Interface-based structural prediction for novel host-pathogen interactions	Pathogens hijack host proteins at multiple levels to circumvent host defence: sequence, structure, motif, and binding surface, or interface.	[96]
14.	Protein-protein interface mimicry	Simulated accessible conformations and various protein-protein interaction (PPI) interfaces have been investigated.	[97]
15.	Host-microbiota interaction networks	Host-microbiota interactions occurred at the molecular level is critical since it could lead to new insights into infection mechanisms.	[14]
16.	Potential molecular signatures via host-microbe interactions	Text mining, filtering, host-microbe interaction, metabolic network reconstruction, simulation, docking, and molecular mimicry studies applied as therapeutic targets.	[67]

(*Continued*)

Table 2.1 Different aspects of recent investigation, interface mimicry, single-cell transcriptomics, manipulations, cell behavior, cell biology, and surface topography, host-microbe interactions. (*Continued*)

S. no.	Various features of cell-material interfaces	Impact of surface chemistry and surface topography on biomaterial-cell interface	Ref.
17.	Responsive cell-material interfaces	Biointerfaces are cell-material interactions, the design of responsive biointerfaces and regenerative medicine initiated complicated biomaterial interfaces that described, including multi-stimulus-responsive surfaces, and interactive biointerfaces.	[22]
18.	Viruses in multiple sclerosis: molecular mimicry	Defined encephalitogenic T-Cell antigen specificity	[63]
19.	Chemistry and material science at cell surface	Tissue engineering, diagnostics, engineering and manipulation of living cells, the basic structural and functional unit of living cells	[98]
20.	Design of biomaterial surfaces: surface modification techniques	Changes in biointerfaces control cell signals that triggered, different biological responses.	[99]
21.	Neuropsychiatric features of COVID-19 and SARS-CoV-2 proteins' mimicry of human protein interactions	Neuropsychiatric symptoms, computational methodology, molecular paths of COVID-19 associated neuropsychiatric symptoms, mimicry of the human protein interactions with SARS-CoV-2 proteins.	[12]

segment to amiloride-sensitive epithelial sodium channel (ENaC)-α was inferred [100]. The existence of the virus in the infected cells was confirmed by these processes in the cells. SARS-CoV-2 produced lung damage by mimicking ENaC in lung epithelial cells [101]. Thus, the inhibition of α-subunit (ENaC-α) of the epithelial sodium channel initiated severe respiratory symptoms or any other events that are currently lacking more investigation.

2.12 Vasculogenic Mimicry and Tumor Angiogenesis

Tumors need a blood supply to develop and spread. Angiogenesis' job during this stage was to recruit new vessels from pre-existent vessels where malignancies were present. The event when genetically deregulated, aggressive tumor cells form microvascular channels is defined as vasculogenic mimicry [102]. Therefore, the vasculogenic mimicry formed cutaneous melanoma and angiogenesis in the neighboring ulceration or necrosis (Figure 2.16). Thus, vascularization, types of vasculogenic mimicry to the cellular section, and angiogenesis occurred in prognostic cutaneous melanoma dermal cells, all of which are recognized as key steps in the regression response [103].

As a result, endothelial cell targeting is highlighted as a therapeutic technique effective in tumor cells without angiogenesis [105]. Furthermore, the biomechanical and molecular mechanisms that control vasculogenic mimicry provide new avenues for designing innovative formulations to treat cancers and associated targets [106]. Besides it, unseen patterning features of vasculogenic mimicry help in designing noninvasive imaging techniques that can be applied for detecting metastases (Table 2.2) [107]. When endothelial cells are present in a tumor with vasculogenic mimicry, an endothelial cell marker is produced, creating it difficult to link vascular

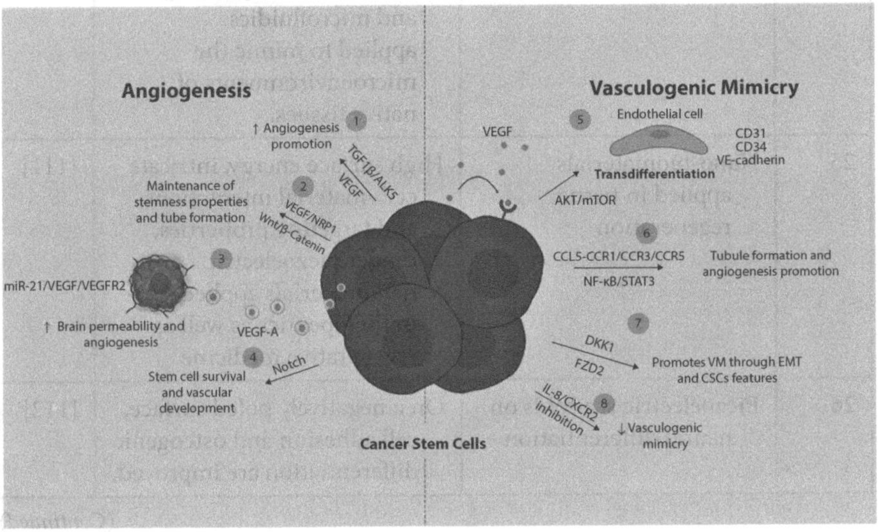

Figure 2.16 VEGF is primarily responsible for CSCs promoting angiogenic and vasculogenic mimics in a variety of cancers. Reprinted (adapted) with permission from [104]. Copyright (2020); Maribel Aguilar-Medina *et al.*, Front. Oncol, Frontiers.

Table 2.2 Different features of mimicry, cell biology, and nanomaterials.

S. no.	Recent findings on molecular mimicry	Advanced materials/ cellular systems for disease diagnosis/tissue regeneration	Ref.
22.	Functionally graded biomaterials and replacement tissues	Engineer functionally graded materials for modelling and repair of diverse tissues, as well as an explanation of how cells interact with different gradients.	[108]
23.	Bioprinting tissue analogues, regeneration and tissue models of cartilage and intervertebral discs	Recapitulation of the native tissues' hierarchical architecture, zone-specific matrix composition, and anisotropic biomechanical behavior.	[109]
24.	Innovations of tissue models of human brain physiology and diseases	Emerged for the design of 3D culture systems developed and applied via bioengineered materials, bioreactors, bioprinting, and microfluidics applied to mimic the microenvironments of native tissues.	[110]
25.	Nano-biomaterials applied in tissue regeneration	High surface energy, intricate cell–material interactions, and targeting properties, render piezoelectric nanomaterials applied in therapeutics as well as regenerative medicine	[111]
26.	Piezoelectric scaffolds on neural differentiation	On a negatively poled surface, cell adhesion and osteogenic differentiation are improved.	[112]

(*Continued*)

Table 2.2 Different features of mimicry, cell biology, and nanomaterials. (*Continued*)

S. no.	Recent findings on molecular mimicry	Advanced materials/ cellular systems for disease diagnosis/tissue regeneration	Ref.
27.	Tendon and collagen and biomechanical applications	Human fibroblasts develop continuously along the axis of bundles, mimicking tendon hierarchy and nanofiber alignment.	[113]
28.	Ligament and tendon repair and regeneration	Biomimetic microenvironment	[114]
29.	Human mesenchymal stem cells for skeletal muscle regeneration	Skeletal muscle mimetic microenvironment, myogenic cell sheets and muscle regeneration upon injection	[115]
30.	Designing scaffolds for corneal regeneration	Human fibroblasts develop continuously along the axis of bundles, mimicking tendon hierarchy and nanofiber alignment.	[116]
31.	Engineering cell-derived matrices and advanced personalized therapies	By mimicking native tissues, cell-derived matrices have emerged to explore cellular processes in tissue-like structures, model disease development, and test tissue regeneration approaches.	[117]
32.	Cell-derived extracellular matrix-rich proliferation, and differentiation	Softer surface and a high concentration of growth factors and related compounds, allowing podocytes to display molecules that control their structure and function.	[118]

(*Continued*)

Table 2.2 Different features of mimicry, cell biology, and nanomaterials. (*Continued*)

S. no.	Recent findings on molecular mimicry	Advanced materials/ cellular systems for disease diagnosis/tissue regeneration	Ref.
33.	Exploring and engineering the cell surface interface	Nanoscale chemistry and topography patterns have been developed to affect cell function using nanoscale material surface engineering.	[119]
34.	Molecular mimicry, host-microbe coevolution	Eukaryotic-like proteins in bacteria are an excellent example of molecular mimicry.	[120]
35.	Molecular mimicry detection and manipulate the maize cell cytoskeleton	Utilize to influence the host metabolism and carry out an effective infection is molecular mimicry.	[121]
36.	Native-mimicking and tumor modeling and tissue engineering	Various three-dimensional (3D) tissue culture systems for native-mimicking *in vitro* milieu were used to mimic native tissue microenvironment.	[122]

(*Continued*)

Table 2.2 Different features of mimicry, cell biology, and nanomaterials. (*Continued*)

S. no.	Recent findings on molecular mimicry	Advanced materials/ cellular systems for disease diagnosis/tissue regeneration	Ref.
37.	Immunoengineering for preventing COVID-19 and infectious diseases	Engineered biomaterials can enable targeted medication delivery, controlled drug release, local immunomodulation, increased immunity, and other desirable functionalities to coordinate proper immune responses and repair tissues, from the nanoscale to the macroscale. Preventive vaccination, immunotherapeutic techniques, therapeutic immunomodulation, biomaterial approaches, and tissue healing and regeneration are highlighted.	[123]
38.	Host-guided viral protein structural mimicry and points to determinants of human disease	Host protein structural mimicry is a common technique used by bacteria. The pathophysiology of COVID-19 can be deduced from knowledge of viral mimics.	[74]

(*Continued*)

Table 2.2 Different features of mimicry, cell biology, and nanomaterials. (*Continued*)

S. no.	Recent findings on molecular mimicry	Advanced materials/ cellular systems for disease diagnosis/tissue regeneration	Ref.
39.	A materials-science perspective and tackling COVID-19	Material platforms for the detection of viruses and the administration of antiviral medications and vaccines, as well as microfluidic devices for in-depth and real-time analysis of viral structures and transmission.	[124]
40.	Motif mimicry incorporating host and pathogen protein interactions	Pathogens use molecular mimicry, host and pathogen proteins share a common host interactor. Similar domain/motif form a mimicry pair with global structural similarity.	[125]
41.	Microfluidics towards mimic blood flow	Identification of pathogenic alterations in red blood cells, white blood cells, and platelets with unprecedented sensitivity, as well as the efficacy assessment of pharmacological treatment, are among the applications.	[126]
42.	Multifaceted cell mimicry	Synthetic cells and how they've been used to a growing number of biologically important phenomena.	[127]
43.	Mimicking microparticles and stem cell science	Regenerative medicine and diagnostic applications, cell imitating microparticles promise to reduce, if not eliminate, the need for living cells.	[128]

(*Continued*)

Table 2.2 Different features of mimicry, cell biology, and nanomaterials. (*Continued*)

S. no.	Recent findings on molecular mimicry	Advanced materials/ cellular systems for disease diagnosis/tissue regeneration	Ref.
44.	Cell-targeted to mimic potent T cell inhibitors	Investigate localization at the target cell–T cell interface, inhibition of proximal TCR signalling, and T cell function suppression.	[129]
45.	Viral molecular, and mimicry influences the antitumor immune response	Autoimmunity is by molecular mimicry have been identified using a bioinformatic method.	[130]

density to angiogenesis. Furthermore, in the tumor cell section, the putative biomechanical and proteolytic features that generate vasculogenic mimicry modify differently.

Vasculogenic mimicry is used as a strong diagnostic of tumor growth since it shows a destructive cell phenotype *in vitro*. These events can be detected using angiography to image vasculogenic mimicry configurations or contemporary ultrasonography to image tumor remodelling. These findings explore the presence of vasculogenic mimicry as metastatic deposits. Therefore, specifically patterned microcirculation exposes metastases [102]. As a result, clinical imaging of vasculogenic mimicry is used to detect melanoma [131]. By injecting indocyanine green into melanomas, a laser scanning can confocal ophthalmoscopy instruments and revealed the existence of loops and networks. For recognized tumors, angiogenic stromal responses and non-angiogenic processes are noted. Traditional tumor "vascularity" markers have been employed as foretelling markers for creating anti-cancer medicines that target angiogenesis in some anti-cancer medications [132]. To target the molecular mechanisms that initiate vasculogenic mimicry and to control the normal physiological processes, vasculogenesis, and vasculogenic mimicry must be detected in advance.

References

1. Stebbins, C.E. and Galaán, J.E., Structural mimicry in bacterial virulence. *Nature*, 412, 6848, 2001.

2. Guven-Maiorov, E., Keskin, O., Gursoy, A., VanWaes, C., Chen, Z., Tsai, C.J., Nussinov, R., TRAF3 signaling: Competitive binding and evolvability of adaptive viral molecular mimicry. *Biochim. Biophys. Acta - Gen. Subj.*, 1860, 11, 2016.
3. Belogurov, A., Kuzina, E., Kudriaeva, A., Kononikhin, A., Kovalchuk, S., Surina, Y., Smirnov, I., Lomakin, Y., Bacheva, A., Stepanov, A., Karpova, Y., Lyupina, Y., Kharybin, O., Melamed, D., Ponomarenko, N., Sharova, N., Nikolaev, E., Gabibov, A., Ubiquitin-independent proteosomal degradation of myelin basic protein contributes to development of neurodegenerative autoimmunity. *FASEB J.*, 29, 1901, 2015.
4. Shahine, A., Van Rhijn, I., Cheng, T.Y., Iwany, S., Gras, S., Branch Moody, D., Rossjohn, J., A molecular basis of human T cell receptor autoreactivity toward self-phospholipids. *Sci. Immunol.*, 2, 16, 2017.
5. Nottingham, U., Cell-materials interface, *Funct. Biointerfaces based Lab. Biophys. Surf. Anal. (LBSA)*, 2022.
6. Berger, G., Escape of pathogens from the host immune response by mutations and mimicry. Possible means to improve vaccine performance. *Med. Hypotheses*, 85, 5, 664–669, 2015.
7. Carter, C., Bioinformatics analysis of homologies between pathogen antigens, autoantigens and the CFTR cystic fibrosis protein: A role for immunoadsorption therapy? *Nat. Preced.*, 586, 1–56, 2010.
8. Santos, E., Orive, G., M, R., Luis, J., Cell-biomaterial interaction: Strategies to mimic the extracellular matrix, in: *On Biomimetics*, 2011.
9. Kumar, R., Chhikara, B.S., Gulia, K., Chhillar, M., Cleaning the molecular machinery of cellsviaproteostasis, proteolysis and endocytosis selectively, effectively, and precisely: Intracellular self-defense and cellular perturbations. *Mol. Omics*, 17, 11, 2021.
10. Santoro, F., Zhao, W., Joubert, L.M., Duan, L., Schnitker, J., Van De Burgt, Y., Lou, H.Y., Liu, B., Salleo, A., Cui, L., Cui, Y., Cui, B., Revealing the cell-material interface with nanometer resolution by focused ion beam/scanning electron microscopy. *ACS Nano*, 11, 8320, 2017.
11. Guven-Maiorov, E., Tsai, C.J., Nussinov, R., Oncoviruses can drive cancer by rewiring signaling pathways through interface mimicry. *Front. Oncol.*, 9, 1236, 2019.
12. Yapici-Eser, H., Koroglu, Y.E., Oztop-Cakmak, O., Keskin, O., Gursoy, A., Gursoy-Ozdemir, Y., Neuropsychiatric symptoms of COVID-19 explained by SARS-CoV-2 proteins' mimicry of human protein interactions. *Front. Hum. Neurosci.*, 15, 656313, 2021.
13. Kumar, R. and Sandeep Mittan, J.S., Nanobiomaterials, nanobiomechanics and tissue bioengineering for advanced regenerative therapeutics: Present and future perceptions. *J. Mater. Nanosci.*, 2, 15, 2015.
14. Jo, J., Abdi Nansa, S., Kim, D.H., Molecular regulators of cellular mechano-adaptation at cell–material interfaces. *Front. Bioeng. Biotechnol.*, 8, 608569, 2020.

15. Guven-Maiorov, E., Tsai, C.J., Nussinov, R., Structural host-microbiota interaction networks. *PLoS Comput. Biol.*, 13, 1, 2017.
16. Xue, M. and Jackson, C.J., Extracellular matrix reorganization during wound healing and its impact on abnormal scarring. *Adv. Wound Care*, 4, 119, 2015.
17. Malonne, H., Langer, I., Kiss, R., Atassi, G., Mechanisms of tumor angiogenesis and therapeutic implications: Angiogenesis inhibitors. *Clin. Exp. Metastasis*, 17, 1, 1999.
18. Nawaz, M., Shah, N., Zanetti, B., Maugeri, M., Silvestre, R., Fatima, F., Neder, L., Valadi, H., Extracellular vesicles and matrix remodeling enzymes: The emerging roles in extracellular matrix remodeling, progression of diseases and tissue repair. *Cells*, 7, 167, 2018.
19. van Weerd, J., Karperien, M., Jonkheijm, P., Supported lipid bilayers for the generation of dynamic cell-material interfaces. *Adv. Healthc. Mater.*, 4, 2743, 2015.
20. Llopis-Hernández, V., Rico, P., Ballester-Beltrán, J., Moratal, D., Salmerón-Sánchez, M., Role of surface chemistry in protein remodeling at the cell-material interface. *PLoS One*, 6, e19610, 2011.
21. Almodovar, J., Castilla Casadiego, D.A., Ramos Avilez, H.V., Polysaccharide-based biomaterials for cell-material interface, in: *Cell Mater. Interface Adv. Tissue Eng. Biosensor, Implant. Imaging Technol.*, vol. 5, p. 230, 2015.
22. Winkler, J., Abisoye-Ogunniyan, A., Metcalf, K.J., Werb, Z., Concepts of extracellular matrix remodelling in tumour progression and metastasis. *Nat. Commun.*, 11, 5120, 2020.
23. Dhowre, H.S., Rajput, S., Russell, N.A., Zelzer, M., Responsive cell-material interfaces. *Nanomedicine*, 10, 849, 2015.
24. Eto, H., Soga, N., Franquelim, H.G., Glock, P., Khmelinskaia, A., Kai, L., Heymann, M., Noji, H., Schwille, P., Design of sealable custom-shaped cell mimicries based on self-assembled monolayers on CYTOP polymer. *ACS Appl. Mater. Interfaces*, 11, 21372, 2019.
25. Li, J., Xun, K., Zheng, L., Peng, X., Qiu, L., Tan, W., DNA-based dynamic mimicry of membrane proteins for programming adaptive cellular interactions. *J. Am. Chem. Soc.*, 143, 4585, 2021.
26. Arias, P., Bellmann, C., Aucouturier, J.J., Facial mimicry in the congenitally blind. *Curr. Biol.*, 31, PR1112, 2021.
27. Carolin Frank, A., Molecular host mimicry and manipulation in bacterial symbionts. *FEMS Microbiol. Lett.*, 366, 1–10, 2019.
28. Liu, J., Radja, A., Gao, Y., Yin, R., Sweeney, A., Yang, S., Mimicry of a biophysical pathway leads to diverse pollen-like surface patterns. *Proc. Natl. Acad. Sci. U. S. A.*, 117, 9699, 2020.
29. Nicolas, J., Magli, S., Rabbachin, L., Sampaolesi, S., Nicotra, F., Russo, L., 3D extracellular matrix mimics: Fundamental concepts and role of materials chemistry to influence stem cell fate. *Biomacromolecules*, 21, 1968, 2020.
30. Lezcano, C., Kleffel, S., Lee, N., Larson, A.R., Zhan, Q., Do Rosario, A., Wang, L.C., Schatton, T., Murphy, G.F., Merkel cell carcinoma expresses

vasculogenic mimicry: Demonstration in patients and experimental manipulation in xenografts. *Lab. Invest.*, 94, 1092, 2014.
31. Hayward, S.D., Liu, J., Fujimuro, M., Notch and Wnt signaling: Mimicry and manipulation by gamma herpesviruses. *Sci. STKE*, 2006.
32. Singh, K.K., Chaubey, G., Chen, J.Y., Suravajhala, P., Decoding sars-cov-2 hijacking of host mitochondria in covid-19 pathogenesis. *Am. J. Physiol. - Cell Physiol.*, 319, C258, 2020.
33. Erisken, C., Kalyon, D.M., Wang, H., Viscoelastic and biomechanical properties of osteochondral tissue constructs generated from graded polycaprolactone and beta-tricalcium phosphate composites. *J. Biomech. Eng.*, 132, 091013, 2010.
34. Unlu, G., Levic, D.S., Melville, D.B., Knapik, E.W., Trafficking mechanisms of extracellular matrix macromolecules: Insights from vertebrate development and human diseases. *Int. J. Biochem. Cell Biol.*, 47, 57, 2004.
35. Valdés-Mora, F., Salomon, R., Gloss, B.S., Law, A.M.K., Venhuizen, J., Castillo, L., Murphy, K.J., Magenau, A., Papanicolaou, M., Rodriguez de la Fuente, L., Roden, D.L., Colino-Sanguino, Y., Kikhtyak, Z., Farbehi, N., Conway, J.R.W., Sikta, N., Oakes, S.R., Cox, T.R., O'Donoghue, S.I., Timpson, P., Ormandy, C.J., Gallego-Ortega, D., Single-cell transcriptomics reveals involution mimicry during the specification of the basal breast cancer subtype. *Cell Rep.*, 35, 108945, 2021.
36. Ramirez, F. and Rifkin, D.B., Cell signaling events: A view from the matrix. *Matrix Biol.*, 22, 101, 2003.
37. Cusick, M.F., Libbey, J.E., Fujinami, R.S., Molecular mimicry as a mechanism of autoimmune disease. *Clin. Rev. Allergy Immunol.*, 42, 102, 2012.
38. Alvarez-Navarro, C., Cragnolini, J.J., Dos Santos, H.G., Barnea, E., Admon, A., Morreale, A., López De Castro, J.A., Novel HLA-B27-restricted epitopes from chlamydia trachomatis generated upon endogenous processing of bacterial proteins suggest a role of molecular mimicry in reactive arthritis. *J. Biol. Chem.*, 288, 25810, 2021.
39. Root-Bernstein, R., Autoimmunity and the microbiome: T-cell receptor mimicry of "self" and microbial antigens mediates self tolerance in holobionts: The concepts of "holoimmunity" (TcR-mediated tolerance for the holobiont) and "holoautoimmunity" (loss of tolerance for the ho. *BioEssays*, 38, 1068, 2016.
40. Rajiv, K., Nanodrugs to end viral infection in pancreatic β-cell and pyknosis, in: *Determ. Nanomedicine Nanotechnol*, vol. 1, p. 28, 2020.
41. Wucherpfennig, K.W. and Strominger, J.L., Molecular mimicry in T cell-mediated autoimmunity: Viral peptides activate human T cell clones specific for myelin basic protein. *Cell*, 80, 695, 1995.
42. De la Herrán-Arita, A.K. and García-García, F., Narcolepsy as an immune-mediated disease. *Sleep Disord.*, 2014, 792687, 2014.
43. Wucherpfenning, K.W. and Strominger, J.L., Molecular mimicry in T cell-mediated autoimmunity: Viral peptides activate human T cell clones

specific for myelin basic protein. *Cell*, 80, 695–705, *Trends Genet.*, 11, 222, 1995.
44. Kumar, N. and Kumar, R., *Nanotechnology and nanomaterials in the treatment of life-threatening diseases*, Elsevier, USA, 2013.
45. Petrova, G., Ferrante, A., Gorski, J., Cross-reactivity of T cells and its role in the immune system. *Crit. Rev. Immunol.*, 32, 349, 2012.
46. Segal, Y. and Shoenfeld, Y., Vaccine-induced autoimmunity: The role of molecular mimicry and immune crossreaction. *Cell. Mol. Immunol.*, 15, 586, 2018.
47. Mertgen, A.S., Trossmann, V.T., Guex, A.G., Maniura-Weber, K., Scheibel, T., Rottmar, M., Multifunctional biomaterials: Combining material modification strategies for engineering of cell-contacting surfaces. *ACS Appl. Mater. Interfaces*, 12, 21342, 2020.
48. González-García, C., Cantini, M., Moratal, D., Altankov, G., Salmerón-Sánchez, M., Vitronectin alters fibronectin organization at the cell-material interface. *Colloids Surf. B Biointerfaces*, 111, 618, 2013.
49. Sarkar, D., Vemula, P.K., Teo, G.S.L., Spelke, D., Karnik, R., Wee, L.Y., Karp, J.M., Chemical engineering of mesenchymal stem cells to induce a cell rolling response. *Bioconjug. Chem.*, 19, 2105, 2008.
50. Li, J., Di Russo, J., Hua, X., Chu, Z., Spatz, J.P., Wei, Q., Surface immobilized E-cadherin mimetic peptide regulates the adhesion and clustering of epithelial cells. *Adv. Healthc. Mater.*, 8, e801384, 2019.
51. Dasgupta, A., Computational challenges for reliability assessment of next-generation micro & nano systems. *7th Int. Conf. Therm. Mech. Multiphysics Simul. Exp. Micro-Electronics Micro-Systems, EuroSimE*, 2006.
52. Bar-Cohen, Y., Biomimetic composites, in: *Biomimetics*, p. 189, CRC Press, UK, 2020.
53. Wang, H., Li, B., Kim, Y.J., Kwon, O.H., Granick, S., Intermediate states of molecular self-assembly from liquid-cell electron microscopy. *Proc. Natl. Acad. Sci. U. S. A.*, 117, 1283, 2020.
54. Zhao, X., Jia, M.X., Jiang, X.K., Wu, L.Z., Li, Z.T., Chen, G.J., Zipper-featured δ-peptide foldamers driven by donor-acceptor interaction. Design, synthesis, and characterization. *J. Org. Chem.*, 69, 270, 2004.
55. Liu, X., Wang, Y., He, Y., Wang, X., Zhang, R., Bachhuka, A., Madathiparambil Visalakshan, R., Feng, Q., Vasilev, K., Synergistic effect of surface chemistry and surface topography gradient on osteogenic/adipogenic differentiation of hMSCs. *ACS Appl. Mater. Interfaces*, 13, 30306, 2021.
56. Łagiewka, J., Girek, T., Ciesielski, W., Cyclodextrins-peptides/proteins conjugates: Synthesis, properties and applications. *Polym. (Basel)*, 13, 1759, 2021.
57. Raddadi, N., Giacomucci, L., Totaro, G., Fava, F., Marinobacter sp. from marine sediments produce highly stable surface-active agents for combatting marine oil spills. *Microb. Cell Factories*, 16, 186, 2017.

58. Nanson, J.D., Rahaman, M.H., Ve, T., Kobe, B., Regulation of signaling by cooperative assembly formation in mammalian innate immunity signalosomes by molecular mimics. *Semin. Cell Dev. Biol.*, 99, 96, 2020.
59. O'Brien, C.M., Holmes, B., Faucett, S., Zhang, L.G., Three-dimensional printing of nanomaterial scaffolds for complex tissue regeneration. *Tissue Eng. - Part B Rev.*, 21, 103, 2015.
60. Ventre, M., Causa, F., Netti, P.A., Determinants of cell-material crosstalk at the interface: Towards engineering of cell instructive materials. *J. R. Soc. Interface*, 9, 2017, 2012.
61. Eisthen, H.L. and Theis, K.R., Animal-microbe interactions and the evolution of nervous systems. *Philos. Trans. R. Soc. B Biol. Sci.*, 371, 1685, 2016.
62. Guven-Maiorov, E., Hakouz, A., Valjevac, S., Keskin, O., Tsai, C.J., Gursoy, A., Nussinov, R., HMI-PRED: A web server for structural prediction of host-microbe interactions based on interface mimicry. *J. Mol. Biol.*, 432, 11, 2020.
63. Geginat, J., Paroni, M., Pagani, M., Galimberti, D., De Francesco, R., Scarpini, E., Abrignani, S., The enigmatic role of viruses in multiple sclerosis: Molecular mimicry or disturbed immune surveillance? *Trends Immunol.*, 38, 7, 2017.
64. Zheng, D., Liwinski, T., Elinav, E., Interaction between microbiota and immunity in health and disease. *Cell Res.*, 30, 6, 2020.
65. Khorsand, B., Savadi, A., Naghibzadeh, M., Comprehensive host-pathogen protein-protein interaction network analysis. *BMC Bioinf.*, 21, 1, 2020.
66. Kowal, C. and Diamond, B., Mimicry between DNA, carbohydrates, and peptides: Implications in systemic lupus erythematosus, in: *Molecular Mimicry, Microbes, and Autoimmunity*, Wiley Online Library, USA, 2014.
67. Verma, A., Sharda, S., Rathi, B., Somvanshi, P., Pandey, B.D., Elucidating potential molecular signatures through host-microbe interactions for reactive arthritis and inflammatory bowel disease using combinatorial approach. *Sci. Rep.*, 10, 1, 2020.
68. Kumar, R. and Chhikara, B.S., Organometallic assemblies: π-electron delocalization, μ-bridging spacers, flexibility, lipophilic nature, bio-accessibility, bioavailability, intracellular trafficking pathways and antimicrobial assimilation. *J. Organomet. Chem.*, 776, 64, 2015.
69. Rauti, R., Shahoha, M., Leichtmann-Bardoogo, Y., Nasser, R., Paz, E., Tamir, R., Miller, V., Babich, T., Shaked, K., Ehrlich, A., Ioannidis, K., Nahmias, Y., Sharan, R., Ashery, U., Maoz, B.M., Effect of SARS-CoV-2 proteins on vascular permeability. *Elife*, 10, e69314, 2021.
70. Beaudoin, C.A., Hamaia, S.W., Huang, C.L.H., Blundell, T.L., Jackson, A.P., Can the SARS-CoV-2 spike protein bind integrins independent of the RGD sequence? *Front. Cell. Infect. Microbiol.*, 11, 765300, 2021.
71. Borbulevych, O.Y., Piepenbrink, K.H., Baker, B.M., Conformational melding permits a conserved binding geometry in TCR recognition of foreign and self molecular mimics. *J. Immunol.*, 186, 2950, 2011.

72. Qin, S., Liu, Y., Tempel, W., Eram, M.S., Bian, C., Liu, K., Senisterra, G., Crombet, L., Vedadi, M., Min, J., Structural basis for histone mimicry and hijacking of host proteins by influenza virus protein NS1. *Nat. Commun.*, 5, 3952, 2014.
73. Bartleson, J.M., Radenkovic, D., Covarrubias, A.J., Furman, D., Winer, D.A., Verdin, E., SARS-CoV-2, COVID-19 and the aging immune system. *Nat. Aging*, 1, 9, 2021.
74. Lasso, G., Honig, B., Shapira, S.D., A sweep of earth's virome reveals host-guided viral protein structural mimicry and points to determinants of human disease. *Cell Syst.*, 12, 1, 2021.
75. Namaka, M., Kapoor, S., Simms, L., Leong, C., Grossberndt, A., Prouta, M., Frost, E., Esfahani, F., Gomori, A., Mulvey, M.R., Molecular mimicry and multiple sclerosis. *Neural Regen. Res.*, 6, 17, 2011.
76. Gran, B., Hemmer, B., Vergelli, M., McFarland, H.F., Martin, R., Molecular mimicry and multiple sclerosis: Degenerate T-cell recognition and the induction of autoimmunity. *Ann. Neurol.*, 45, 5, 559, 1999.
77. Stebbins, C.E. and Galán, J.E., Modulation of host signaling by a bacterial mimic: Structure of the Salmonella effector SptP bound to Rac1. *Mol. Cell*, 6, 6, 2020.
78. Bathawab, F., Bennett, M., Cantini, M., Reboud, J., Dalby, M.J., Salmerón-Sánchez, M., Lateral chain length in polyalkyl acrylates determines the mobility of fibronectin at the cell/material interface. *Langmuir*, 32, 800, 2016.
79. Christen, U., Hintermann, E., Holdener, M., von Herrath, M.G., Viral triggers for autoimmunity: Is the "glass of molecular mimicry" half full or half empty? *J. Autoimmun.*, 34, 38, 2010.
80. Qin, S., Liu, Y., Tempel, W., Eram, M., Bian, C., Liu, K., Senisterra, G., Crombet, L., Vedadi, M., Histone mimicry and hijacking of host proteins by influenza virus protein NS1. *Acta Crystallogr. Sect. A Found. Adv.*, 70, C1591, 2014.
81. Alcami, A., Viral mimicry of cytokines, chemokines and their receptors. *Nat. Rev. Immunol.*, 3, 36, 2003.
82. Wu, D., Wang, L., Zhang, Y., Bai, L., Yu, F., Emerging roles of pathogen-secreted host mimics in plant disease development. *Trends Parasitol.*, 37, 1082, 2021.
83. Kori, M. and Arga, K.Y., Pathways involved in viral oncogenesis: New perspectives from virus-host protein interactomics. *Biochim. Biophys. Acta - Mol. Basis Dis.*, 1866, 165885, 2020.
84. Alto, N.M. and Orth, K., Subversion of cell signaling by pathogens. *Cold Spring Harb. Perspect. Biol.*, 4, a06114, 2012.
85. Beckstead, B.L., Santosa, D.M., Giachelli, C.M., Mimicking cell-cell interactions at the biomaterial-cell interface for control of stem cell differentiation. *J. Biomed. Mater. Res. - Part A*, 79, 94, 2006.

86. Anselme, K., Ploux, L., Ponche, A., Cell/material interfaces: Influence of surface chemistry and surface topography on cell adhesion. *J. Adhes. Sci. Technol.*, 24, 831, 2010.
87. Vrana, N.E., *Cell and material interface advances in tissue engineering, biosensor, implant, and imaging technologies*, Routledge Taylor and Francis Group, UK, 2016.
88. Mukwaya, V., Mann, S., Dou, H., Chemical communication at the synthetic cell/living cell interface. *Commun. Chem.*, 4, 161, 2021.
89. Chan, S.L., Low, L.Y., Hsu, S., Li, S., Liu, T., Santelli, E., Le Negrate, G., Reed, J.C., Woods, V.L., Pascual, J., Molecular mimicry in innate immunity. *J. Biol. Chem.*, 284, 21386, 2009.
90. Asano, S.M., Gao, R., Wassie, A.T., Tillberg, P.W., Chen, F., Boyden, E.S., Expansion microscopy: Protocols for imaging proteins and RNA in cells and tissues. *Curr. Protoc. Cell Biol.*, 80, e56, 2018.
91. Belardi, B., Son, S., Felce, J.H., Dustin, M.L., Fletcher, D.A., Cell–cell interfaces as specialized compartments directing cell function. *Nat. Rev. Mol. Cell Biol.*, 21, 12, 2020.
92. Mano, J.F. and Choi, I.S., Advanced control over cell-material interfaces. *Polym. (Basel)*, 9, 704, 2017.
93. Wechman, S.L., Emdad, L., Sarkar, D., Das, S.K., Fisher, P.B., Vascular mimicry: Triggers, molecular interactions and *in vivo* models, in: *Advances in Cancer Research*, vol. 148, p. 27, 2020.
94. Qian, X., Nymann Westensee, I., Brodszkij, E., Städler, B., Cell mimicry as a bottom-up strategy for hierarchical engineering of nature-inspired entities. *Wiley Interdiscip. Rev. Nanomed. Nanobiotechnol.*, 13, e1683, 2021.
95. Mallet, J., Mimicry: An interface between psychology and evolution. *Proc. Natl. Acad. Sci. U. S. A.*, 98, 8928, 2021.
96. Guven-Maiorov, E., Tsai, C.J., Ma, B., Nussinov, R., Interface-based structural prediction of novel host-pathogen interactions. *Methods Mol. Biol.*, 1851, 317, 2019.
97. Li, X., Taechalertpaisarn, J., Xin, D., Burgess, K., Protein-protein interface mimicry by an oxazoline piperidine-2,4-dione. *Org. Lett.*, 17, 632, 2015.
98. Zhao, W., Teo, G.S.L., Kumar, N., Karp, J.M., Chemistry and material science at the cell surface. *Mater. Today*, 13, 14, 2010.
99. Amani, H., Arzaghi, H., Bayandori, M., Dezfuli, A.S., Pazoki-Toroudi, H., Shafiee, A., Moradi, L., Controlling cell behavior through the design of biomaterial surfaces: A focus on surface modification techniques. *Adv. Mater. Interfaces*, 6, 1900572, 2019.
100. Anand, P., Puranik, A., Aravamudan, M., Venkatakrishnan, A.J., Soundararajan, V., SARS-CoV-2 strategically mimics proteolytic activation of human ENaC. *Elife*, 9, e58603, 2020.
101. Karoyan, P., Vieillard, V., Gómez-Morales, L., Odile, E., Guihot, A., Luyt, C.E., Denis, A., Grondin, P., Lequin, O., Human ACE2 peptide-mimics block SARS-CoV-2 pulmonary cells infection. *Commun. Biol.*, 4, 97, 2021.

102. Fernández-Cortés, M., Delgado-Bellido, D., Javier Oliver, F., Vasculogenic mimicry: Become an endothelial cell "But not so much." *Front. Oncol.*, 9, 803, 2019.
103. Andonegui-Elguera, M.A., Alfaro-Mora, Y., Cáceres-Gutiérrez, R., Caro-Sánchez, C.H.S., Herrera, L.A., Díaz-Chávez, J., An overview of vasculogenic mimicry in breast cancer. *Front. Oncol.*, 10, 220, 2020.
104. Lizárraga-Verdugo, E., Avendaño-Félix, M., Bermúdez, M., Ramos-Payán, R., Pérez-Plasencia, C., Aguilar-Medina, M., Cancer stem cells and its role in angiogenesis and vasculogenic mimicry in gastrointestinal cancers. *Front. Oncol.*, 10, 413, 2020.
105. El-Fiqi, A., Mandakhbayar, N., Jo, S.B., Knowles, J.C., Lee, J.H., Kim, H.W., Nanotherapeutics for regeneration of degenerated tissue infected by bacteria through the multiple delivery of bioactive ions and growth factor with antibacterial/angiogenic and osteogenic/odontogenic capacity. *Bioact. Mater.*, 6, 123, 2021.
106. Folberg, R. and Maniotis, A.J., Vasculogenic mimicry. *APMIS*, 112, 508, 2004.
107. Luo, Q., Wang, J., Zhao, W., Peng, Z., Liu, X., Li, B., Zhang, H., Shan, B., Zhang, C., Duan, C., Vasculogenic mimicry in carcinogenesis and clinical applications. *J. Hematol. Oncol.*, 13, 19, 2020.
108. Lowen, J.M. and Leach, J.K., Functionally graded biomaterials for use as model systems and replacement tissues. *Adv. Funct. Mater.*, 30, 909089, 2020.
109. Vernengo, A.J., Grad, S., Eglin, D., Alini, M., Li, Z., Bioprinting tissue analogues with decellularized extracellular matrix bioink for regeneration and tissue models of cartilage and intervertebral discs. *Adv. Funct. Mater.*, 30, 1909044, 2020.
110. Lovett, M.L., Nieland, T.J.F., Dingle, Y.T.L., Kaplan, D.L., Innovations in 3D tissue models of human brain physiology and diseases. *Adv. Funct. Mater.*, 30, 1909146, 2020.
111. Kapat, K., Shubhra, Q.T.H., Zhou, M., Leeuwenburgh, S., Piezoelectric nano-biomaterials for biomedicine and tissue regeneration. *Adv. Funct. Mater.*, 30, 1909045, 2020.
112. Lee, Y.S. and Arinzeh, T.L., The influence of piezoelectric scaffolds on neural differentiation of human neural stem/progenitor cells. *Tissue Eng. - Part A*, 18, 19, 2012.
113. Sensini, A., Gualandi, C., Zucchelli, A., Boyle, L.A., Kao, A.P., Reilly, G.C., Tozzi, G., Cristofolini, L., Focarete, M.L., Tendon fascicle-inspired nanofibrous scaffold of polylactic acid/collagen with enhanced 3D-structure and biomechanical properties. *Sci. Rep.*, 8, 1, 2018.
114. Maghdouri-White, Y., Petrova, S., Sori, N., Polk, S., Wriggers, H., Ogle, R., Ogle, R., Francis, M., Electrospun silk-collagen scaffolds and BMP-13 for ligament and tendon repair and regeneration. *Biomed. Phys. Eng. Express*, 4, 025013, 2018.

115. Yoon, J.K., Misra, M., Yu, S.J., Kim, H.Y., Bhang, S.H., Song, S.Y., Lee, J.R., Ryu, S., Choo, Y.W., Jeong, G.J., Kwon, S.P., Im, S.G., Lee, T.I., Kim, B.S., Thermosensitive, stretchable, and piezoelectric substrate for generation of myogenic cell sheet fragments from human mesenchymal stem cells for skeletal muscle regeneration. *Adv. Funct. Mater.*, 27, 1703853, 2017.
116. Ahearne, M., Fernández-Pérez, J., Masterton, S., Madden, P.W., Bhattacharjee, P., Designing scaffolds for corneal regeneration. *Adv. Funct. Mater.*, 30, 1908996, 2020.
117. Rubi-Sans, G., Castaño, O., Cano, I., Mateos-Timoneda, M.A., Perez-Amodio, S., Engel, E., Engineering cell-derived matrices: From 3D models to advanced personalized therapies. *Adv. Funct. Mater.*, 30, 2000496, 2020.
118. Satyam, A., Tsokos, M.G., Tresback, J.S., Zeugolis, D.I., Tsokos, G.C., Cell-derived extracellular matrix-rich biomimetic substrate supports podocyte proliferation, differentiation, and maintenance of native phenotype. *Adv. Funct. Mater.*, 30, 44, 2020.
119. Stevens, M.M. and George, J.H., Exploring and engineering the cell surface interface. *Science*, 310, 5751, 2005.
120. Mondino, S., Schmidt, S., Buchrieser, C., Molecular mimicry: A paradigm of host-microbe coevolution illustrated by legionella. *MBio*, 11, e01201, 2020.
121. Armijos-Jaramillo, V., Espinosa, N., Vizcaıno, K., Santander-Gordon, D., A novel in silico method for molecular mimicry detection finds a formin with the potential to manipulate the maize cell cytoskeleton. *Mol. Plant-Microbe Interact.*, 34, 815, 2021.
122. Rijal, G. and Li, W., Native-mimicking *in vitro* microenvironment: An elusive and seductive future for tumor modeling and tissue engineering. *J. Biol. Eng.*, 12, 20, 2018.
123. Zarubova, J., Zhang, X., Hoffman, T., Hasani-Sadrabadi, M.M., Li, S., Biomaterial-based immunoengineering to fight COVID-19 and infectious diseases. *Matter*, 4, 1528, 2021.
124. Tang, Z., Kong, N., Zhang, X., Liu, Y., Hu, P., Mou, S., Liljeström, P., Shi, J., Tan, W., Kim, J.S., Cao, Y., Langer, R., Leong, K.W., Farokhzad, O.C., Tao, W., A materials-science perspective on tackling COVID-19. *Nat. Rev. Mater.*, 5, 847, 2020.
125. Tayal, S., Bhatia, V., Mehrotra, T., S.B., ImitateDB: A database for domain and motif mimicry incorporating host and pathogen protein interactions. *Amino Acids*, 1, 1, 2020.
126. Sebastian, B. and Dittrich, P.S., Microfluidics to mimic blood flow in health and disease. *Annu. Rev. Fluid Mech.*, 50, 483, 2018.
127. Mason, A.F. and van Hest, J.C.M., Multifaceted cell mimicry in coacervate-based synthetic cells. *Emerg. Top. Life Sci.*, 3, 567, 2019.
128. Labriola, N.R., Azagury, A., Gutierrez, R., Mathiowitz, E., Darling, E.M., Concise review: Fabrication, customization, and application of cell mimicking microparticles in stem cell science. *Stem Cells Transl. Med.*, 7, 232, 2018.

129. Curnock, A.P., Bossi, G., Kumaran, J., Bawden, L.J., Figueiredo, R., Tawar, R., Wiseman, K., Henderson, E., Hoong, S.J., Gonzalez, V., Ghadbane, H., Knight, D.E.O., O'Dwyer, R., Overton, D.X., Lucato, C.M., Smith, N.M.G., Reis, C.R., Page, K., Whaley, L.M., McCully, M.L., Hearty, S., Mahon, T.M., Weber, P., Cell-targeted PD-1 agonists that mimic PD-L1 are potent T cell inhibitors. *JCI Insight*, 6, e152468, 2021.
130. Chiaro, J., Kasanen, H.H.E., Whalley, T., Capasso, C., Grönholm, M., Feola, S., Peltonen, K., Hamdan, F., Hernberg, M., Mäkelä, S., Karhapää, H., Brown, P.E., Martins, B., Fusciello, M., Ylösmäki, E.O., Greco, D., Kreutzman, A.S., Mustjoki, S., Szomolay, B., Cerullo, V., Viral molecular mimicry influences the antitumor immune response in murine and human melanoma. *Cancer Immunol. Res.*, 9, 981, 2021.
131. Luo, Q., Wang, J., Zhao, W., Peng, Z., Liu, X., Li, B., Zhang, H., Shan, B., Zhang, C., Duan, C., Vasculogenic mimicry in carcinogenesis and clinical applications. *J. Hematol. Oncol.*, 13, 19, 2020.
132. Wehland, M., Bauer, J., Magnusson, N.E., Infanger, M., Grimm, D., Biomarkers for anti-angiogenic therapy in cancer. *Int. J. Mol. Sci.*, 14, 9338, 2013.

3

Bacteriocins of Lactic Acid Bacteria as a Potential Antimicrobial Peptide

Ajay Kumar[1], Rohit Ruhal[2,3] and Rashmi Kataria[1*]

[1]*Department of Biotechnology, Delhi Technological University, Shahbad Daulatpur, New Delhi, India*

[2]*Department of Biological Sciences, Marquette University, Milwaukee, Wisconsin, United States*

[3]*School of Bioscience and Technology (SBST), Vellore Institute of Technology, Vellore, Tamil Nadu, India*

Abstract

Antibiotic overuse accelerated a rise in its resistance against bacteria. Lactic acid-producing bacteria generate antibacterial peptides called bacteriocins. They can increase shelf life naturally in various food products by serving as a bacteria-killing action. Lactic acid bacteria are Gram-negative, mesophilic, fastidious bacteria and considered as food-grade bacteria that produce hydrogen peroxide, organic acids, diacyls and antimicrobial compounds. Lactic acid bacteria peptides are gaining popularity as potential therapeutic antimicrobials to replace antibiotics. Based on physicochemical and structural features, they are divided into three groups. To combat the rise of epidemics associated with foodborne harmful microbes, the food sector has invested much in research. For their antibacterial qualities, many bioactive beneficial substances are derived from *Lactobacillus* and in food industries for food preservation.

Keywords: Bacteriocin, peptides, antimicrobial activity, food industry, lactic acid bacteria

3.1 Introduction

Antibiotic usage for microbial infection treatment in humans, as well as in animals causes the generation of antibiotic resistance in bacteria [1].

**Corresponding author*: rashmikataria@dtu.ac.in; rashmikataria@gmail.com

Inamuddin, Tariq Altalhi and Ashjan Alrogi (eds.) Biomimicry Materials and Applications, (83–104)

Antimicrobial drug resistance has increased dramatically because of using these resulting in the spread of resistant microbes to people. Antibiotic-resistant microorganisms have created a threat to global human food and health in the few decades [2]. This is because of antibiotic misuse, and poor ingestion, led to the rapid occurrence of multi-drug-resistant microbes [3]. Antibiotics can influence the gut microbiome and host metabolism by limiting pathogen invasion, altering the function of immune function, boosting fat absorption by minimizing conjugated bile salt hydrolysis and improving nutrient uptake [4]. Pathogenic bacteria can have an impact on animal health along with their development and reproductive potential. Some of these may make their way into the human digestive tract via the food chain [5, 6]. Antimicrobial resistance causes an increment in dependency on the healthcare system, costs of morbidity, mortality, and hospitalization and is raising the concern [7].

Several multidrug-resistant microbes are responsible for creating infections that are challenging to treat. Pathogens from the *Acinetobacter*, *Pseudomonas*, *Enterobacteriaceae*, and *Pseudomonas* genera have already developed resistance to several antibiotics, which are extremely powerful antibiotics frequently used as life-saving drugs [8].

Bacteria synthesize a significant concentration of bioactive peptides having antimicrobial properties against several food pathogens, these are known as Bacteriocin. These proteaceous metabolites are composed of bactericidal activity which restricts the growth of related species [9]. Bacteriocins generated from gram-positive bacteria, on the other hand, have a somewhat larger range of antimicrobial action. These proteinaceous molecules are destroyed in the human digestive system by protease digestion; as a result, they become inert in the human intestine and have no adverse impact. Food producers believe the production of bacteriocins to be advantageous since they are effective at killing or confining harmful bacteria that compete for the same environment in adequate concentrations [10].

Because of the sluggish pace of development of novel antimicrobials having a broad spectrum, researchers discovered a viable alternative by harnessing compounds produced by good microbes to kill pathogenic bacteria microbes to minimize the emergence of multidrug resistance. Bacteriocins are one such antibiotic-free solution [11]. Bacteriocins have a broad range of uses due to their antibacterial capabilities, including medicinal, esthetic, and food packaging materials to increase life span [12].

New methods of food preservation are urgently needed due to the increasing demand for food items that include the least concentrations of artificial additives, it has become necessary to investigate novel methods of food preservation. Bacteriocins, which are antibiotics, have lately received a great deal of interest in this area. Because of the existence of self-protective proteins, they have been shown to have the capacity to inhibit the development of germs that cause food deterioration without inflicting damage to them. Lactic acid-producing bacteria are reported to produce bacteriocins, which are non-toxic and safe for intake and prevent harmful bacterial growth [13]. The utilization of bacterial bioactive peptides as natural food additives in different food sectors has been investigated, and it has been shown that they may extend the life span of minimally least processed food items by acting as a killing machine on the bacteria. They attack and disrupt the cell plasma membrane, prevent the development of unwanted bacteria [14], and lead to the production of the pores in the target bacterial cell membrane. This chapter gives an overview of bacteriocins, their Classification, mode of action on target bacteria, and applications in various industries.

Lactic acid-producing bacteria also known as LAB are being used as a natural preservative for fermented foods due to their ability to impart notable flavor and taste improvements while inhibiting food degrading microorganisms [15]. LAB are accepted as food-grade bacteria and are hence considered safe for human consumption [16].

Bacteriocins are heat resistant and vulnerable to protein deactivation [17, 18]. Genes for bacteriocin biosynthesis have been discovered on plasmids and chromosomes [19, 20], which are responsible for bacteriocin production.

3.2 Bacteriocins

Bacteriocins synthesized in ribosomes are small antimicrobial bioactive peptides, Induced by an extra post-translational modification. These are strong enough to kill taxonomically similar species even in very small quantities [21]. During the initial period of formation, it is the main metabolite [24]. For the most part, they are only effective against germs that are strongly linked to them. Bacteriocins are of 2 to 30 kDa in size. These peptides are small chain molecules, prepared of up to 60 amino acids, hydrophilic in nature, and polar helices able to withstand temperatures up to 100°C for 10 minutes [22, 23]. Their range of activity, manner of action, molecular mass, evolutionary origin, and metabolic features varies [25].

They are garnering a lot of attention because they are safe for public health can be consumed without any negative effect and fall under the category of generally recognized as safe (GRAS) [26]. Furthermore, these are synthesized in ribosome and their activity range may be increased by changing their properties [27]. Bacteriocins may be an efficient treatment due to their wide range of activity against antibiotic drug-resistant bacteria [28].

3.3 Lactic Acid Bacteria

LAB are gram-negative, mesophilic, negative for catalase test, acid-tolerant, fastidious bacteria which have low GC content. They are non-spore-forming, facultative anaerobes, Which can utilize a range of sugars for fermentation. They may be present in a variety of food products, such as meat, dairy, vegetables, fruits, grains, and various kinds of traditional fermented cuisines [29]. The LAB group is now categorized as part of the Firmicutes phylum, Bacilli class, and Latobacillales order [30]. Table 3.1 represents the major species of lactic acid bacteria that produce bacteriocins. Cell appearance, sugar fermentation mode, growth thermal gradient, and sugar utilization patterns are used to classify LAB [31]. These are one of the most significant bacteria utilized in fermentation processes and for enriching the flavor and texture of fermented foods [32]. These microbes synthesize lactic acid as the primary product from different sugar and side products

Table 3.1 Major genera and species of bacteriocin producing Lactic acid bacteria.

Genera	Species	References
Lactobacillus	*Lactobacillus plantarum TF711, L. plantarum, L. fermentum, L. acidophilus, Lactobacillus sake 148.*	[31]
Lactococcus	*Lactococcus lactis lactis, Lactococcus lactis cremoris*	[31]
Enterococcus	*Enterococcus avium LMGT 3465, Enterococcus faecium RC714, Enterococcus faecium P13, Enterococcus hirae DCH5*	[32]
Pediococcus	*P. acidilactici, P. pentosaceus and P. damnosus*	[33]
Streptococcus	*Streptococcus, S. salivarius, S. pyogenes, S. macedonicus, S. thermophilus, S. mutans, S. uberis, S. bovis, S. rattuss*	[34]
Carnobacterium	*Carnobacterium piscicola, Carnobacterium divergens*	[35]
Leuconostoc	*Leuconostoc mesenteroides, Leuconostoc citreum GJ7, Leuconostoc gelidum*	[36]

are usually bacteriocins, hydrogen peroxide, and other compounds that impede the expansion of food degrading pathogens [33].

3.4 Classification of LAB Bacteriocins

Bacteriocins are a large group of chemicals produced by LAB [34] (Table 3.1). The classification of bacteriocin is majorly based on the number of amino acids, properties, antimicrobial activity against pathogens, and the mechanism of action. In past years, various classifications have been proposed and reviewed [35, 36]. In one of the classifications, the basis was taken as the type of bacteriocin production in Gram-negative or Gram-positive bacteria [37, 38]. A more recently reviewed classification system of LAB is given by Alvarez-Sieiro and it is being currently being used.

3.4.1 Class I Bacteriocins or Lantibiotics

Lantibiotics which fall under this class are very small molecules with a size of below 5 kDa and composed of a chain of 19 to 38 amino acids, heat resistant. These further go for posttranslational modifications and form the peptides with polycyclic thioether amino acids including lanthionine, methyl-lanthionine, and 2-amino isobutyric acid [39]. These Lantibiotics are further divided into 5 classes, based on charge [40]. Nisin is the most significant bacteriocin in this class. Nisin targets cell wall and membrane of pathogen and causes pore formation. This results in the outflow of essential elements, ATP molecules, ions, and amino acids through the pores. This changes the permeability and ultimately causes cell death of pathogens [41].

3.4.1.1 Class Ia

In this class, the size is less than 10 kDa [42]. It includes all peptides modified by different enzymes and give different structure and amino acid composition. These are composed of peptide part which is important for enzyme identification, and transportation, and keep it inactive. When it is fused with a core peptide it performs the activity including the formation of pore and lysis of the cell membrane [43, 44]. The bacteriocins including nisin and lacticin are long and screw-shaped. These are positively charged molecules of the very small size of about 2 to 4 kDa. The mechanism of action includes the formation of a pore in the pathogen's cell membrane and causes degradation of the cytoplasmic membrane [45].

3.4.1.2 *Class Ib*

The bacteriocins of this class are known as head-to-tail cyclized peptides. These are the bacteriocins of less than 10 kDa. These do not need an enzyme for activation or maturation. Only a leader peptidase and transporter are required for functioning [23]. It is composed of globular structures [14], with a negative net charge. These enzymes perform as inhibitors and are very sensitive in terms of immunological reaction. These molecules interfere in important cellular metabolic reactions like cell wall synthesis. They act by disturbing the important enzymatic reactions in pathogenic microbes [14]. Mersacidin is another type of cyclized compound that is secreted by *Bacillus* spp. [46, 47].

3.4.1.3 *Class Ic or Antibiotics*

Sactipeptides or antibiotics are compounds that are composed of antimicrobial activity. These are sulphur-to-α-carbon-containing peptide molecules [48]. There is no report of the presence of sactipeptide in LAB so far and only prediction studies have been performed. Researchers are working in this field and more potential results are awaiting [23].

3.4.1.4 *Class Id*

These are another class of peptide molecules with linear azol (in) e-containing peptides (LAPs), these are composed of different types of cyclic compounds like heterocyclic rings of thiazole and methyl oxazole. These are formed from different amino acids cysteine, serine, and threonine [49]. The most significant LAP produced by LAB is streptolysin S. The mechanism of its action is still not identified. Another molecule, microcin, synthesized in *E. coli* also comes into this class.

3.4.1.5 *Class Ie*

Bacteriocins molecules of this class are also known as glycocins, which are composed of glycosylated amino acid residues. Glycocin F is discovered in LAB, *Lactobacillus plantarum* in this class [50]. In glycocin F the amino acid residues are arranged in two alpha helices and bonded by disulfide bonding [23]. The mode of action of glycocins is also not clear. Other than glycocin F, another bacteriocin, enterocin F4-9 was identified from *Enterococcus faecium* [51], and just like glycocin F, it is anticipated to be bacteriostatic [50].

3.4.1.6 *Class If*

In this class, peptide bonding is present between the first amino acid residues in the core peptide and is also known as lasso peptides. The negative charge is present at position +7 to +9 and form a ring [52]. There is no report of lasso peptides form. Lasso peptides are known to have various variety functional properties, including antibacterial, antiviral, and anticancer, which help in various pathogenic conditions [53].

3.4.2 Class II Bacteriocins

These class II-bacteriocins are also known as post-translationally modified small peptides and composed of diversed types of ribosomally antimicrobial molecules [54]. These are also small peptides molecules of size less than 10 kDa. Class II-bacteriocins are stable at high temperature and present in an amphiphilic helical form. These molecules can enter in pathogenic cell cytoplasm, and this cause depolarization and cause ultimately cell death [55].

3.4.3 Class III Bacteriocins

It is a class of molecule with unmodified bacteriocins molecules with higher molecular size (>30 kDa). These molecules are also thermotolerant and these peptides are of multi domain and have various catalytic property. This class is again subdivided in two subclasses: subclass IIIa also known as bacteriolysins and subclass IIIb known as Bacteriolysins. These are large polypeptide molecules with a size of 27 to 35 kDa. They are very susceptible to rising temperature levels and cause disruption of cellular structure of bacteria. Bacteriolysins is composed of two portions: the N-terminal catalytic chain and the C-terminal detection chain and They are linked together by a linker helix. The catalytic portion is characterized as a hydrolytic protease which disrupt the peptides and peptidoglycan cross-links [56]. The most well-known bacteriolysins produced by Lactic acid bacteria are enterolysin A (*Enterococcus faecalis LMG 2333*) [57, 58], lysostafin (*Staphylococcus simulans*) [59], millericin B (*Streptococcus milleri NMSCC 061*) [37, 56].

3.5 Mechanisms of LAB Bacteriocins to Inactivate Microbial Growth

A Belgian scientist André Gratia developed and reported first plasmid-derived antimicrobial peptide compounds [59]. Bacteriocin's antibiotic

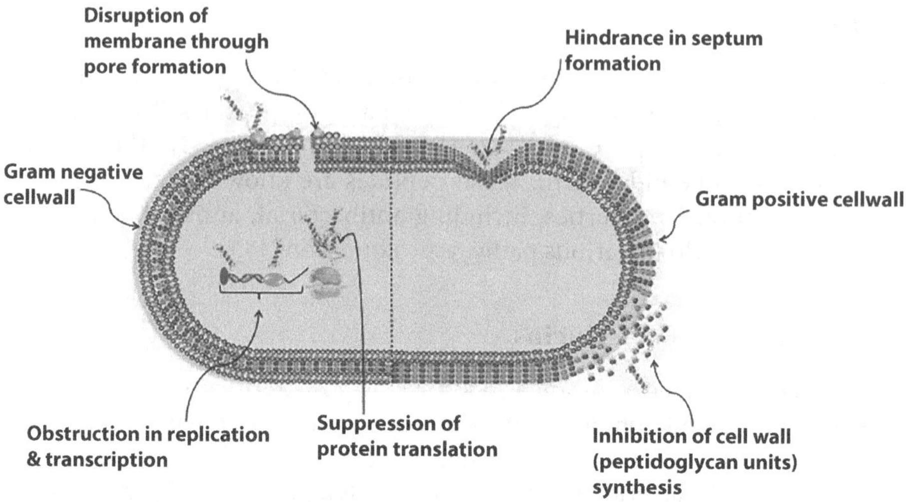

Figure 3.1 Mechanisms of action of bacteriocins produced by lactic acid bacteria.

composed of some characteristics like inhibition in cell membrane synthesis and deteriorate internal important cell building blocks like DNA and RNA in non-host pathogenic strains (Figure 3.1) [60]. Different mechanism for inactivation are listed in Table 3.2.

3.5.1 Action on Cell Wall Synthesis

3.5.1.1 Pore Formation

A bacteriocin Lactococcin, destroys sensitive food pathogen cells by creating cellular membrane holes that are dependent on the existence of a receptor site of protein. The hole formation causes membrane integrity in sensitive cells to depolarize, resulting in reduction the capacity of amino acid uptake and outflow of the absorbed amino acids. The class IIa bacteriocin molecules have been shown to create pores in specific cells [61, 62]. Ubericin K kills cells by disrupting their membranes by the carbohydrate transporter as a targeted site [63]. Several molecules, such as nisin, subtilin, lacticin, and streptococcin, establish the pores formation in the plasma membrane of target microbial cell. It is a multistep system that works with the bacteriocin's interplay with the membrane and ends with its incorporation in the lipid bilayers, secretion of low molecular weight molecules, such as amino acid residues, anion, cations, and adenosine triphosphate [50].

Table 3.2 Lactic acid bacteria bacteriocins and their bactericidal mechanism of actions

Producer strain	Bacteriocin	Mechanism of action
Lactococcus Lactis	Nisin A	Pore Formation
Enterococcus Faecalis	Cytolysin L1	Pore Formation
Staphylococcus epidermis	Pep5	Pore Formation
Carnobacteriumpiscicola	Carnocin U149	Pore Formation
Bacillus subtilis	Subtilin	Pore Formation
Lactobacillus sake	Lactocin S	Pore Formation
Staphylococcus epidermis	Epilancin K7	Pore Formation
Streptococcus mutans	Mutacin	Pore Formation
Staphylococcus epidermis	Epidermin	Pore Formation
Streptococcus salivarius	Salivaricin	Pore Formation
Staphylococcus gallinarum	Gallidermin	Pore Formation
Streptococcus pyogenes	Streptococcin	Membrane permeabilization
Lactococcus lactis	Lacticin 481	Inhibits PBP1b-catalysed peptidoglycan formation
Actinoplanes sp.	Actagardine	Inhibition of cell wall formation
Bacillus sp.	Mersacidin	Inhibition of cell wall formation
Streptococcus cinnamoneus	Duramycin	Membrane disintegration

3.5.1.2 Inhibition of Peptidoglycan Synthesis

Lactococcus lactis produces Nisin A, a bacteriocin that hinders the formation of the cell wall. Lipid 2 is a cell wall synthesis precursor that creates pores in the membrane and leads to cell death [64]. When used in high concentrations, the main mechanism is bacteriostatic, while the secondary mechanism is bactericidal. Type B lantibiotics are considered globular peptides involving mersacidin, duramycin, mersacidin, and actagardin. They restrict the cell wall synthesis by disrupting the enzyme functionality [65].

Lipid II–mediated mechanism of action is indicated by antibiotics that links to lipid II and destroys the bacteria without impermealizing their cell membranes [66]. Type-B lantibiotics mersacidin and actagardin, which have fundamentally different structures than nisin, obstruct the transglycosylation reaction during peptidoglycan production by creating a complex with lipid II [50]. Mersacidin inhibits the addition of glucose and D-alanine in cell wall of bacteria, which causes the ceasing of its production and results in cell death in the *Staphylococcus simulans* 22 strain [67].

3.5.2 Obstruction in Replication and Transcription

There is a difference in the structure of human and bacterial DNA gyrase, making it an excellent antibacterial target [68]. Microcins, which are mostly generated by *Enterobacteriaceae*, are thermostable, pH, and protease-resistant. Microcins kill bacteria by several processes, comprising hole creation, nuclease activities like DNase and RNase, and suppression of protein biosynthesis or DNA synthesis [69]. LAB peptides that are not bactercidal (class IIIb) function by interfering with cell glucose absorption, starving the cells, and altering membrane gradient potential. Another approach is to block aim bacteria from producing protein molecules [40].

3.5.3 Inhibition in Protein Synthesis

Bacteriocins, particularly colicins, limit the synthesis of proteins by interacting with the protein translation process at several stages [70]. Bacteriocins with rRNase activity at 16S include colicins E3-E6, cloacin, and DF13 fall under this catagory. Endonuclease activity in Group E colicins inhibits translation by cleaving 16S rRNA and acting on the 30 end of the coding sequence [27]. Certain colicins, such as E5, are tRNases proteins that work by causing tRNA depletion, resulting in restricted protein production [71]. *Enterobacter cloacae* DF 13 produces a bacteriocin DF13 which strongly inhibit *Klebsiella edwardsiivar.edwardsii* growth [72]. Botromycins, which are generated by bacteria classified under the genus *Streptomyces* spp., are identified to suppress protein synthesis by interfering with the bacterial 50S ribosome.

3.5.4 Disruption of Membrane Structure

Bacteriocins like nisin they exploit a lipid flip flop to create toroidal holes, allowing protein spillage and other cellular components to escape, resulting in cell death. Similarly, carnocycline binds to the lipid bilayer directly and forms ion-specific membrane holes [27], while plantaricin S [52] and leuconocin S damage the bacterial cell membrane [51]. Microcins interact with a variety of cellular targets, resulting in several mechanisms of action which destruct membrane. It affects enzymatic system of cell such as inhibition of ATP synthase, inhibition of RNA polymerase, degradation of DNA and aspartyl-t RNA Furthermore, colicin synthesis is fatal to the synthesizing microbial strain since its release is linked to membrane disruption and ultimately leads to cell death [53, 54].

3.5.5 Disruption in Septum Formation

Whenever food pathogens observe unfavorable conditions around them, there is an induction in the formation of a septum, which results in spore formation. Septum development prevention is a novel strategy of biopeptides that happens throughout the mitosis division of cells [27]. Garvicin A, a newly discovered linear peptide class IId bacteriocin synthesised from *Lactococcus garvieae*, inhibits the formation of septum in *L. garvieae* cells. Its unique inhibitory spectrum, as well as its capacity to block septum formation, could be applied to cure the infection instigated by pathogenic microbial strains of *L. garvieae* strains [55].

3.6 Antimicrobial Properties of LAB Bacteriocins

Bacteriocin and organic acid are the most often cited mechanisms for LAB's antibacterial properties. Bacteriocin has gained attention as a technique for food bio-preservation and as a bioweapon against pathogenic microbes [56]. Bacteriocins are the most potential substitute for antimicrobials. In the food sector, lactic acid bacterial antimicrobial peptides could be a replacement for conventional preservatives [56]. To kill bacteria, bacteriocins work by attaching to receptors on their surfaces and then penetrating the membrane, resulting in cytotoxicity and ultimately resulting in bacterial death. Bacteriocins are also low-toxic peptides that are vulnerable to proteases, like pepsin and trypsin [72].

3.6.1 Antiviral Activity

Several studies have indicated that Bacteriocins may be utilized as antiviral agents. A different class of bacteriocins demonstrated antiviral activity against the herpes simplex virus [39, 57]. It is also demonstrated by Wachsman *et al.* that, enterocins CRL35 and ST4V, which are produced by *Enterococcus faecium*, showed antiviral activities against HSV-1 and HSV-2. These affect intracellular viral proliferation and inhibit the final step of replication [58]. *Actinomadura namibiensis* produces Labyrintho peptin A1 [59]. LabyA1 functions as an antiviral against HIV and HSV infection and their transmission [60]. It has a unique carbon cyclic post-translationally modified labionin residue. LabyA1 was discovered to prevent HIV-infected T cells from transmitting the virus to uninfected CD4 + T cells [61].

3.6.2 Antibacterial Properties

The growing emergence of multidrug-resistant infectious diseases has prompted a quest for alternate means of infection eradication [62]. Bacterial peptides have a high antibacterial capability toward drug-resistant bacteria [63]. They have a different mechanism of action when opposed to traditional drugs, which may reduce the possibility of cross-resistance formation, making them viable antimicrobial substitutes [64]. Even at a very small dosage in picomolar, they demonstrate significant efficacy and selectivity toward microorganisms [65].

Enterocin AS-48 a cyclic peptide [66, 67] is a potent antimicrobial compound against *Staphylococci* [68]. Mutacin 1140 is produced by *Streptococcus mutans* [69] and salivaricin D produced by *Streptococcus salivarius.* Nisin F has high inhibitory activities against *P. aeruginosa*, *Streptococcus pneumonia*, and *S. aureus*. Pediocin PA-1, is found successfully prevents *L. monocytogenes* infection [70] and subtilosin A synthesized by *Bacillus subtilis* [71] inhibits *Gardnerella vaginalis.*

3.6.3 Antifungal Activity

Bacteriocins have been shown to have the ability to inhibit fungal food deterioration [72, 73]. *Aspergillus flavus*, *A. parasiticus*, *A. fumigatus*, *Penicillium roqueforti*, *Fusarium verticillioides*, and *F. graminearum* produce bacteriocin-like inhibitory substance (BLIS) as a potential antifungal against *Aspergillus*, *Penicillium*, and *Fusarium* genera [74]. Bacteriocin released by *Enterococcus faecalis* has been found to reduce virulence and biofilm formation in *Candida albicans* strains, it suppresses the hyphae production and retard the growth [75]. LAB bacteriocins' antifungal action has not been adequately documented in the literature.

3.7 Applications

3.7.1 Bacteriocins in Packaging Film

Bacteriopeptides are usually absorbed in food products and break down quickly, leading to a loss of antimicrobial properties [76], including the bacterial proteins in the preservation of less integrated food products by utilizing the bacteriocins [77] is an alternate strategy. Coverings for food packing that promote bioactivity in complicated food supply chains [78] is known as active wrapping, which, in contrast to shielding the goods from the outside atmosphere, also prevents microbiological development, reducing the need for additives to be added directly. Because it has fewer ingredients, it also meets the desire for nutritious products [79]. Nisin/methylcellulose covering is utilized on polyethylene films, and nisin coating is often used on chicken [80].

3.7.2 Potential Use as Biopreservatives

Biopreservation techniques utilize bioactive compounds originating from microorganisms, such as bacteria, fungus, herbs, or creatures to increase the storage life span of food while ensuring its quality. Such chemicals may come from various sources, including bacteria, plants, livestock dairy products, or eggs and animals [38]. Bacteriocins offer a lot of promise in the food business for extending the shelf life of many foodstuffs, either single or in a conjunction with the other storage technologies, which is termed as hurdle technology [81]. Because aggressive microbes in fruit products are becoming more common in foodstuff, it is necessary to look at alternative techniques for increasing the standards and security of these products. Biopeptides from lactic acid bacteria might be employed as an added barrier to suppress harmful and rotting bacteria from growing in fruit [82].

3.7.3 Bacteriocins as Antibiofilm

Biofilm accumulates on synthetic turf such as polyester, glassware, steels, plastic, latex, and wood articles in the food service sectors [81]. Diverse apparatus like butter turbines, unpasteurized milk containers, pasteurizers, cheese containers, tubes, and packaging articles that allow the development of biofilm by reacting surface layer with high or low temperatures [83]. *Lactobacillus lactis* secretes nisin, which prevents the synthesis of cell wall, increasing transmembrane sensitivity [84]. Whenever magnesium and calcium ions are employed in meals, enhancing susceptibility aids in the creation of apertures, allowing them to come out more easily [85]. Bacteriocin's use has expanded to include pharmaceutical techniques for managing persistent multi-drug-resistant bacterial strains, in addition to acting as a natural preservative [86].

3.7.4 Applications in Foods Industries

During cultivation, harvesting, transportation, and handling the agricultural products, feces and environmental components from the field may contaminate fruits and vegetables with deadly bacteria [87]. The fruit products may have a bacterial population of 104–106 cfu gl before they enter the manufacturing sector [88]. This exterior contamination might enter the fruit's inside during manufacturing and end up in the final output. The holding, chopping, shredding, dicing, and slicing steps of the processing are all possible major contaminants that result in an increased microbial count [89]. In fruits such as apple, berries mango, strawberry, grape, and orange beverages, *Bifidobacterium animalis* and *enterocin AS-48* synthesized from *Enterococcus faecalis AS48* [90, 91] restrict the vegetal development and spore formation of *A. acidoterrestris*. The growth of *L. monocytogenes* in different meat products can be reduced by mixing multiple category of IIa bacteriocins. BacFL31, a moderately purified bacteriocin produced by *Enterococcus faecium* sp., affects sensory qualities and the storage of cold minced turkey meat and also inhibits infections, such as *Listeria monocytogenes* and *Salmonella typhimurium*. LAB's bacteriocin preparations may be used as a bioconservative meat product, including readily available meat, and may inhibit dangerous microbes and/or rotting bacteria [92]. Bacteriocins may also be utilized to increase food processing, speed up cheese ripening, and even improve the taste of cheese [93]. Nisin has been frequently used to substitute nitrate in cheese and processed cheese propagates to reduce *clostridia* spore propagation [94], and it is also reported as successful in the treatment of various diseases infected by bacteria, such as *Listeria monocytogenes* and *Staphylococcus aureus* in dairy foods [62]. In newborn milk composition, natural yogurt, and cream cheese, lacticin 3147 powdered formulation was demonstrated as an effective molecule for the control of bacteria such as *Listeria* and *Bacillus* [95].

3.8 Conclusion

Due to the increasing problem of multi-drug-resistant microbes bacteriocins are taking more attention and it is a very important area of current research. Bacteriocins have shown antimicrobial activities against several pathogenic microorganisms. Bacteriocins are demonstrated important applications against food infections. Hence, these are used in food applications as bacteriocins fall under the category of GRAS. Their mechanism of action on the target cell is very important. Still, several classes of these compounds are needed to identify and isolate for various industrial, as well as pharmaceutical applications.

Acknowledgment

The authors would like to thank the financial support given by the Department of Biotechnology (DBT), Govt. of India with grant BT/RLF/Re-entry/40/2017 and SERB project file no: EEQ/2020/000614.

References

1. Mittal, A.K., Bhardwaj, R., Mishra, P., Rajput, S.K., Antimicrobials misuse/overuse: Adverse effect, mechanism, challenges and strategies to combat resistance. *Open Biotechnol. J.*, 14, 1, 2001.
2. Ma, F., Xu, S., Tang, Z., Li, Z., Zhang, L., Use of antimicrobials in food animals and impact of transmission of antimicrobial resistance on humans. *Biosaf. Health*, 3, 1, 32–38, 2021.
3. Ventola, C.L., The antibiotic resistance crisis: Part 1: Causes and threats. *P T: J. Formulary Manage.*, 40, 4, 277–283, 2015.
4. Shah, T., Baloch, Z., Shah, Z., Cui, X., Xia, X., The intestinal microbiota: Impacts of antibiotics therapy, colonization resistance, and diseases. *Int. J. Mol. Sci.*, 22, 12, 6597, 2021.
5. Vieco-Saiz, N., Belguesmia, Y., Raspoet, R., Auclair, E., Gancel, F., Kempf, I., Drider, D., Benefits and inputs from lactic acid bacteria and their bacteriocins as alternatives to antibiotic growth promoters during food-animal production. *Front. Microbiol.*, 10, 57, 2019.
6. Soucy, S.M., Huang, J., Gogarten, J.P., Horizontal gene transfer: Building the web of life. *Nat. Rev. Genet.*, 16, 472–482, 2015.
7. Peters, L., Olson, L., Khu, D., Linnros, S., Le, N.K., Hanberger, H., Hoang, N., Tran, D.M., Larsson, M., Multiple antibiotic resistance as a risk factor for mortality and prolonged hospital stay: A cohort study among neonatal intensive care patients with hospital-acquired infections caused by gram-negative bacteria in Vietnam. *PLoS One*, 14, 5, 2019.
8. Terreni, M. and Marina, M., New antibiotics for multidrug-resistant bacterial strains: Latest research developments and future perspectives. *Mol.*, 26, 2671, 2021.

9. Nazari, M. and Smith, D.L., A PGPR-produced bacteriocin for sustainable agriculture: A review of thuricin 17 characteristics and applications. *Front. Plant Sci.*, 11, 916, 2020.
10. GradisteanuPircalabioru, G., Popa, L.I., Marutescu, L., Gheorghe, I., Popa, M., Czobor Barbu, I., Cristescu, R., Chifiriuc, M.C., Bacteriocins in the era of antibiotic resistance: Rising to the challenge. *Pharceutics*, 13, 2, 196, 2021.
11. Leite, E., Carmo, F.L., Barh, D., Ghosh, P., Berkova, N., Azevedo, V., Junior, A., Bacteriocins as an alternative in the treatment of infections by Staphylococcus aureus. *An. Acad. Bras. Ciênc.*, 92, 17, 2020.
12. Becerril, R., Nerín, C., Silva, F., Encapsulation systems for antimicrobial food packaging components: An update. *Mol.*, 25, 1134, 2020.
13. Simons, A., Alhanout, K., Duval, R.E., Bacteriocins, antimicrobial peptides from bacterial origin: Overview of their biology and their impact against multidrug-resistant bacteria. *Micro.*, 8, 5, 639, 2020.
14. Sidhu, P. and Nehra, K., Bacteriocins of lactic acid bacteria as potent antimicrobial peptides against food pathogens, in: *Biomimetics*, 2021.
15. Bintsis, T., Lactic acid bacteria as starter cultures: An update in their metabolism and genetics. *AIMS Microbiol.*, 4, 4, 665, 2018.
16. Kieliszek, M., Pobiega, K., Piwowarek, K., Kot, A.M., Characteristics of the proteolytic enzymes produced by lactic acid bacteria. *Mol.*, 26, 7, 1858, 2021.
17. Imade, E.E., Omonigho, S.E., Babalola, O.O., Lactic acid bacterial bacteriocins and their bioactive properties against food-associated antibiotic-resistant bacteria. *Ann. Microbiol.*, 71, 44, 2021.
18. Mondal, R. and Aruna, B., Biopreservative activity of bacteriocin-producing lactic acid bacteria isolated from fermented green gram batter. *Int. J. Food Ferment. Technol.*, 4, 121, 2014.
19. Lagha, A.B., Haas, B., Gottschalk, M., Grenier, D., Antimicrobial potential of bacteriocins in poultry and swine production. *Vet. Res.*, 48, 22, 2017.
20. Daba, G.M. and Elkhateeb, W.A., Bacteriocins of lactic acid bacteria as biotechnological tools in food and pharmaceuticals: Current applications and future prospects. *Biocatal. Agric. Biotechnol.*, 28, 101750, 2020.
21. Paweł, M., *Pet-To-Man Travelling Staphylococci*, pp. 161–171, Elsevier BV, 2018.
22. Surekha, M. and Reddy, S.M., Preservatives classification and properties, in: *Encyclopedia of Food Microbiology*, Second Edition, C.A. Batt and M.L. Tortorello (Eds.), pp. 69–75, Academic Press, 2014.
23. Alvarez-Sieiro, P., Montalbán-López, M., Mu, D., Kuipers, O.P., Bacteriocins of lactic acid bacteria: Extending the family. *Appl. Microbiol. Biotechnol.*, 100, 7, 2939, 2016.
24. Gaspar, C., Donders, G.G., Palmeira-de-Oliveira, R., Queiroz, J.A., Tomaz, C., Martinez-de-Oliveira, J., Palmeira-de-Oliveira, A., Bacteriocin production of the probiotic Lactobacillus acidophilus KS400. *AMB Express*, 27, 8, 1, 153, 2018.

25. Mokoena, M.P., Lactic acid bacteria and their bacteriocins: Classification, biosynthesis and applications against uropathogens: A mini-review. *Mol.*, 22, 8, 1255, 2017.
26. Perez, R.H., Novel bacteriocins from lactic acid bacteria (LAB): Various structures and applications. *Microb. Cell Fact.*, 13, S3, 2014.
27. Sulthana, R. and Archer, A.C., Bacteriocin nanoconjugates: Boon to medical and food industry. *J. Appl. Microbiol.*, 131, 1056–1071, 2021.
28. Ayukekbong, J.A., Ntemgwa, M., Atabe, A.N., The threat of antimicrobial resistance in developing countries: Causes and control strategies. *Antimicrob. Resist. Infect. Control*, 6, 47, 2017.
29. Papadimitriou, K., Alegría, Á., Bron, P.A., de Angelis, M., Gobbetti, M., Kleerebezem, M., Lemos, J.A., Linares, D.M., Ross, P., Stanton, C., Turroni, F., van Sinderen, D., Varmanen, P., Ventura, M., Zúñiga, M., Tsakalidou, E., Kok, J., Stress physiology of lactic acid bacteria. *Microbiol. Mol. Biol. Rev.*, 80, 3, 837, 2016.
30. Yanglei, Y., Ping, L., Fan, Z., Tingting, Z., Yuanyuan, S., Xin, W., Bianfang, L., Yujie, C., Xin, Z., Xin, L., Current status and potentiality of class II bacteriocins from lactic acid bacteria: Structure, mode of action and applications in the food industry. *Trends Food Sci. Technol.*, 120, 387, 2022.
31. Parada, J., Bacteriocins from lactic acid bacteria: Purification, properties and use as biopreservatives. *Braz. Arch. Biol. Technol.*, 50, 512, 2022.
32. Kjos, M., Salehian, Z., Diep, D., An extracellular loop of the mannose phosphotransferase system component IIC is responsible for specific targeting by class IIa bacteriocins. *J. Bacteriol.*, 192, 5906, 2010.
33. Papagianni, M. and Anastasiadou, S., The bacteriocins of Pediococci. Sources, production, properties and applications. *Microb. Cell Fact.*, 8, 3, 2009.
34. Nes, I.F., Diep, D.B., Holo, H., Bacteriocin diversity in Streptococcus and Enterococcus. *J. Bacteriol.*, 189, 4, 1189, 2007.
35. Pilet, M.F., Xavier, D., Rachel, B., Georges, N., Michel, D., Jean-C, P., Evidence for two bacteriocins produced by Carnobacterium piscicola and Carnobacterium divergens isolated from fish and active against Listeria monocytogenes. *J. Food Prot.*, 58, 3, 256, 1995.
36. Dicks, L.M.T., Dreyer, L., Smith, C., van Staden, A.D., Corrigendum: A review: The fate of bacteriocins in the human gastro-intestinal tract: Do they cross the gut–blood barrier? *Front. Microbiol.*, 9, 2938, 2018.
37. Afrin, S., Hoque, M.A., Sarker, A.K., Satter, M.A., Bhuiyan, M., Characterization and profiling of bacteriocin-like substances produced by lactic acid bacteria from cheese samples. *Access Microbiol.*, 3, 6, 2021.
38. Ayikpoe, R.S. and van der Donk, W.A., Peptide backbone modifications in lanthipeptides. *Methods Enzymol.*, 656, 573–621, 2021.
39. Pérez-Ramos, A., Madi-Moussa, D., Coucheney, F., Drider, D., Current knowledge of the mode of action and immunity mechanisms of LAB-bacteriocins. *Microorg.*, 9, 10, 2107, 2021.

40. Bin Hafeez, A., Jiang, X., Bergen, P.J., Zhu, Y., Antimicrobial peptides: An update on classifications and databases. *Int. J. Mol. Sci.*, 22, 21, 11691, 2021.
41. Karbalaei-Heidari, H.R. and Budisa, N., Combating antimicrobial resistance with new-to-nature lanthipeptides created by genetic code expansion. *Front. Microbiol.*, 11, 590522, 2021.
42. Arnison, P.G., Bibb, M.J., Bierbaum, G., Bowers, A.A., Bugni, T.S., Bulaj, G., Camarero, J.A., Campopiano, D.J., Challis, G.L., Clardy, J., Cotter, P.D., Craik, D.J., Dawson, M., Dittmann, E., Donadio, S., Dorrestein, P.C., Entian, K.D., Fischbach, M.A., Garavelli, J.S., Göransson, U., van der Donk, W.A., Ribosomally synthesized and post-translationally modified peptide natural products: Overview and recommendations for a universal nomenclature. *Nat. Prod. Rep.*, 30, 1, 108, 2013.
43. Zhao, X., Xu, Y., Viel, J.H., Kuipers, O.P., Semisynthetic macrocyclic lipo-lanthipeptides display antimicrobial activity against bacterial pathogens. *ACS Synth. Biol.*, 10, 8, 2021.
44. Negash, A. and Tsehai, B., Current applications of bacteriocin. *Int. J. Microbiol.*, 1, 4374891, 2020.
45. Gang, D., Kim, D.W., Park, H.S., Cyclic peptides: Promising scaffolds for biopharmaceuticals. *Gene*, 9, 11, 557, 2018.
46. Sumi, C.D., Yang, B.W., Yeo, I.C., Hahm, Y.T., Antimicrobial peptides of the genus Bacillus: A new era for antibiotics. *Can. J. Microbiol.*, 61, 2, 2015.
47. Mathur, H., Field, D., Rea, M.C., Cotter, P.D., Hill, C., Ross, R.P., Bacteriocin-antimicrobial synergy: A medical and food perspective. *Front. Microbiol.*, 8, 1205, 2017.
48. Telhig, S., Ben, S.L., Zirah, S., Fliss, I., Rebuffat, S., Bacteriocins to thwart bacterial resistance in gram negative bacteria. *Front. Microbiol.*, 11, 586433, 2021.
49. Sharma, P., Kaur, S., Chadha, B.S. *et al.*, Anticancer and antimicrobial potential of enterocin 12a from Enterococcus faecium. *BMC Microbiol.*, 21, 39, 2021.
50. Waliczek, M., Wierzbicka, M., Arkuszewski, M., Kijewska, M., Jaremko, Ł., Rajagopal, P., Attempting to synthesize lasso peptides using high pressure. *PLoS One*, 15, 6, 2020.
51. Cheng, C. and Hua, Z.C., Lasso peptides: Heterologous production and potential medical application. *Front. Bioeng. Biotechnol.*, 8, 571165, 2020.
52. Khorshidian, N., Khanniri, E., Mohammadi, M., Mortazavian, A.M., Yousefi, M., Antibacterial activity of pediocin and pediocin-producing bacteria against listeria monocytogenes in meat products. *Front. Microbiol.*, 12, 709959, 2021.
53. Mauriello, G., Control of microbial activity using antimicrobial packaging, in: *Antimicr. Food. Pk.*, pp. 141–152, 2016.
54. Lohans, C.T. and Vederas, J.C., Development of class IIa bacteriocins as therapeutic agents. *Int. J. Microbiol.*, 13pp, Article 386410, 2012.

55. Sibinelli-Sousa, S., Hespanhol, J.T., Bayer-Santos, E., Targeting the Achilles' heel of bacteria: Different mechanisms to break down the peptidoglycan cell wall during bacterial warfare. *J. Bacteriol.*, 203, 7, e00478-20, 2021.
56. El-Gendy, A.O., Brede, D.A., Essam, T.M., Purification and characterization of bacteriocins-like inhibitory substances from food isolated Enterococcus faecalis OS13 with activity against nosocomial enterococci. *Sci. Rep.*, 11, 3795, 2021.
57. Jayakumar, J., Kumar, V., Biswas, L., Biswas, R., Therapeutic applications of lysostaphin against Staphylococcus aureus. *J. Appl. Microbiol.*, 131, 1072, 2021.
58. Moravej, H., Moravej, Z., Yazdanparast, M., Heiat, M., Mirhosseini, A., Moghaddam, M.M., Mirnejad, R., Antimicrobial peptides: Features, action, and their resistance mechanisms in bacteria. *Microb. Drug Resist.*, 24, 6, 747, 2018.
59. Wiedemann, I., Benz, R., Sahl, H.G., Lipid II-mediated pore formation by the peptide antibiotic nisin: A black lipid membrane study. *J. Bacteriol.*, 186, 10, 3259, 2004.
60. Rios, C., Natalia, S., Chalón, M., Navarro, S., Pediocin-like bacteriocins: New perspectives on mechanism of action and immunity. *Curr. Genet.*, 64, 345–351, 2018.
61. Panina, I., Krylov, N., Nolde, D., Environmental and dynamic effects explain how nisin captures membrane-bound lipid II. *Sci. Rep.*, 10, 882, 2020.
62. Shin, J.M., Gwak, J.W., Kamarajan, P., Fenno, J.C., Rickard, A.H., Kapila, Y.L., Biomedical applications of nisin. *J. Appl. Microbiol.*, 120, 6, 1449, 2016.
63. Bengtsson, T., Lönn, J., Khalaf, H., Palm, E., The lantibioticgallidermin acts bactericidal against Staphylococcus epidermidis and Staphylococcus aureus and antagonizes the bacteria-induced proinflammatory responses in dermal fibroblasts. *Microbiologyopen*, 7, 6, 2018.
64. Viel, J.H., van Tilburg, A.Y., Kuipers, O.P., Characterization of leader processing shows that partially processed mersacidin is activated by AprE after export. *Front. Microbiol.*, 12, 765659, 2021.
65. Duprey, A. and Groisman, E.A., DNA supercoiling differences in bacteria result from disparate DNA gyrase activation by polyamines. *PLoS Genet.*, 16, 10, e1009085, 2020.
66. Paquette, S.J. and Reuter, T., Properties of an antimicrobial molecule produced by an *Escherichia coli* champion. *Antibiotics*, 9, 1, 6, 2020.
67. Xing, J., Weston, K., Yong-C, K., Seok, H.H., Rapid production and characterization of antimicrobial colicins using *Escherichia coli*-based cell-free protein synthesis. *Synth. Biol.*, 3, 1, ysy004, 2018.
68. Chopra, S. and Reader, J., tRNAs as antibiotic targets. *Int. J. Mol. Sci.*, 16, 1, 321–349, 2014.
69. de Graaf, F.K., Spanjaerdt, E.A., Stouthamer, A.H., Mode of action of a bacteriocin produced by *Enterobacter cloacae* DF 13. *Anton. Leeuw.*, 35, 287, 969, 2018.

70. Mei, M., Thomas, J., Diggle, S.P., Heterogenous susceptibility to R-pyocins in populations of *Pseudomonas aeruginosa* sourced from cystic fibrosis lungs. *mBio*, 12, 3, e00458-21, 2021.
71. Calcuttawala, F., Pal, A., Nath, P. *et al.*, Structural and functional insights into colicin: A new paradigm in drug discovery. *Arch. Microbiol.*, 204, 37, 2022.
72. Dubey, S., Diep, D.B., Evensen, Ø., Munang'andu, H.M., Garvicin, K.S., A broad-spectrum bacteriocin protects Zebrafish Larvae against Lactococcus garvieae infection. *Int. J. Mol. Sci.*, 23, 5, 2833, 2011.
73. Chen, J., Pang, H., Wang, L., Ma, C., Wu, G., Liu, Y., Guan, Y., Zhang, M., Qin, G., Tan, Z., Bacteriocin-producing lactic acid bacteria strains with antimicrobial activity screened from Bamei pig feces. *Foods*, 11, 5, 709, 2022.
74. Fernandes, A. and Jobby, R., Bacteriocins from lactic acid bacteria and their potential clinical applications. *Appl. Biochem. Biotechnol.*, 11, 51, 2022.
75. Tiwari, S.K., Dicks, L.M.T., Popov, I.V., Karaseva, A., Ermakov, A.M., Suvorov, A., Tagg, J.R., Weeks, R., Chikindas, M.L., Probiotics at war against viruses: What is missing from the picture? *Front. Microbiol.*, 11, 1877, 2020.
76. Blockus, S., Sake, S.M., Wetzke, M., Grethe, C., Graalmann, T., Pils, M., Le Goffic, R., Galloux, M., Prochnow, H., Rox, K., Hüttel, S., Rupcic, Z., Wiegmann, B., Dijkman, R., Rameix-Welti, M.A., Eléouët, J.F., Duprex, W.P., Thiel, V., Hansen, G., Brönstrup, M., Pietschmann, T., Labyrinthopeptins as virolytic inhibitors of respiratory syncytial virus cell entry. *Antivir. Res.*, 177, 104774, 2020.
77. Fu, Y., Jaarsma, A.H., Kuipers, O.P., Antiviral activities and applications of ribosomally synthesized and post-translationally modified peptides (RiPPs). *Cell. Mol. Life Sci.*, 78, 8, 3921, 2021.
78. Férir, G., Petrova, M.I., Andrei, G., Huskens, D., Hoorelbeke, B., Snoeck, R., The lantibiotic peptide labyrinthopeptin A1 demonstrates broad anti-HIV and anti-HSV activity with potential for microbicidal applications. *PLoS One*, 8, 5, e64010, 2013.
79. Roberto, V., Ana, A.T., Silvio, S.D., Jain, S., Multidrug-resistant bacteria and alternative methods to control them: An overview. *Microb. Drug Resist.*, 25, 6, 890–908, 2019.
80. Pontes, J.T., Toledo, B., Roque-B, Pavan, F.R., Antimicrobial peptides as an alternative for the eradication of bacterial biofilms of multi-drug resistant bacteria. *Pharma.*, 14, 3, 642, 2022.
81. Flynn, J., Ryan, A., Hudson, S.P., Pre-formulation and delivery strategies for the development of bacteriocins as next generation antibiotics. *Eur. J. Pharm. Biopharm. :Off. J. Arb. Pharm. Ver.*, 165, 149, 2021.
82. Dai, M., Li, Y., Xu, L., Wu, D., Zhou, Q., Li, P., Gu, Q., A novel bacteriocin from *Lactobacillus Pentosus* ZFM94 and its antibacterial mode of action. *Front. Nutr.*, 8, 710862, 2021.
83. Ross, J.N., Fields, F.R., Kalwajtys, V.R., Gonzalez, A.J., O'Connor, S., Zhang, A., Moran, T.E., Hammers, D.E., Carothers, K.E., Lee, S.W., Synthetic peptide libraries designed from a minimal alpha-helical domain of AS-48-bacteriocin

homologs exhibit potent antibacterial activity. *Front. Microbiol.*, 11, 589666, 2020.
84. Perales, A.J., Rubiño, S., Martínez-B, M., Valdivia, E., Montalbán-L, M., Cebrián, R., Maqueda, M., LAB Bacteriocins Controlling the Food Isolated (Drug-Resistant) *Staphylococci. Front. Microbiol.*, 9, 1143, 2018.
85. Telhig, S., Ben Said, L., Zirah, S., Fliss, I., Rebuffat, S., Bacteriocins to thwart bacterial resistance in gram negative bacteria. *Front. Microbiol.*, 11, 586433, 2020.
86. Kers, J.A., Sharp, R.E., Defusco, A.W., Park, J.H., Xu, J., Pulse, M.E., Weiss, W.J., Handfield, M., Mutacin 1140 lantibiotic variants are efficacious against clostridium difficile infection. *Front. Microbiol.*, 9, 415, 2018.
87. Lawrence, G., Garcia-G, Enriqueta, W., Calum, P., Begley, M., Cotter, P., Guinane, C., Nisin G is a novel nisin variant produced by a gut-derived Streptococcus salivarius, bioRxiv 2022.02.15.480493, 2022.
88. Kuniyoshi, T.M., O'Connor, P.M., Lawton, E., Thapa, D., Mesa-Pereira, B., Abulu, S., Hill, C., Ross, R.P., Oliveira, R., Cotter, P.D., An oxidation resistant pediocin PA-1 derivative and penocin a display effective anti-Listeria activity in a model human gut environment. *Gut Microbes*, 14, 1, 200407, 2021.
89. Epparti, P., Eligar, S., Sattur, A., Kumar, G., Halami, P.M., Characterization of dual bacteriocins producing *Bacillus subtilis* SC3.7 isolated from fermented food. *Lebensmittel Wissenschaft + Technologies*, 154, 112854, 2021.
90. Qian, Z., Zhao, D., Yin, Y., Zhu, H., Chen, D., Antibacterial activity of Lactobacillus strains isolated from Mongolian yogurt against Gardnerella vaginalis. *Biomed. Res. Int.*, 9, Article 1787, 2020.
91. Garcha, S. and Rani, P., Antifungal activity of bacteriocins of Lactobacillus plantarum MTCC 9503 purified using diatomite calcium silicate. *Int. J. Food Ferment. Technol.*, 4, 27, 2014.
92. de Souza de Azevedo, P.O., Mendonça, C.M.N., Moreno, A.C.R., Antibacterial and antifungal activity of crude and freeze-dried bacteriocin-like inhibitory substance produced by Pediococcuspentosaceus. *Sci. Rep.*, 10, 12291, 2020.
93. Oshiro, K.G.N., Rodrigues, G., Monges, B.E.D., Cardoso, M.H., Franco, O.L., Bioactive peptides against fungal biofilms. *Front. Microbiol.*, 10, 2169, 2019.
94. Anvar, A.A., Ahari, H., Ataee, M., Antimicrobial properties of food nanopackaging: A new focus on foodborne pathogens. *Front. Microbiol.*, 12, 690706, 2021.
95. Punia Bangar, S., Chaudhary, V., Thakur, N., Kajla, P., Kumar, M., Trif, M., Natural antimicrobials as additives for edible food packaging applications: A review. *Foods*, 10, 2282, 2021.

4

A Review on Emergence of a Nature-Inspired Polymer-Polydopamine in Biomedicine

Lakshmi Nidhi Rao[1], Arun M. Isloor[2*], Aditya Shetty[1] and Pallavi K.C.[2]

[1]*Department of Conservative Dentistry and Endodontics, A.B. Shetty Memorial Institute of Dental Sciences, Nitte Deemed to be University, Deralakatte, Mangalore, India*

[2]*Membrane and Separation Technology Laboratory, Department of Chemistry, National Institute of Technology Karnataka, Surathkal, Mangalore, India*

Abstract

Biological structures have evolved throughout the millennia. Nature has been proactive with continual improvement to fine-tune the material properties resulting in optimization of the structure–function relationship. In keeping with this trend, mussels' capacity to stick to varied moist surfaces with sufficient strength to endure powerful ocean currents has aroused curiosity and research into the role of polydopamine (PDA). It is a flexible and organic substance that exhibits individual mechanical characteristics and excellent fixation to different substrate stuff in a humidified environment. It is shown to be structurally inert and has been proved to be harmless to living systems. This makes it ideal to be used as a coating material. It has spurred the usage of the substance as molecular glue due to its perfect adherence as seen in mussels. Conformal polydopamine coatings offer sole physical and chemical properties to various substrate stuff, such as polymers, metallic things, ceramics, and many more. This has been found to help accentuate the existing properties of the coated material. PDA is a highly malleable material that may be used as a nanocomposite, a nanoparticle, and as a coating for existing materials. The additional PDA properties in biocompatibility, biodegradability, anti-microbial activity, bone regeneration, and versatility make it a promising material that can be mapped into various fields of biomedicine. In this review, we focus upon key structural aspects and related properties of PDA and how they could potentially hold as a tool for multitude biomedical and dental applications.

**Corresponding author*: isloor@yahoo.com

Inamuddin, Tariq Altalhi and Ashjan Alrogi (eds.) Biomimicry Materials and Applications, (105–126)

This article will go over some recent research on polydopamine advancements in the biomedical domain. The mechanism of polymerization is first discussed followed by the various forms of polydopamine nanostructures, as well as their latest uses in biological disciplines, particularly in drug administration. The review finally is concluded by a summary of the findings.

***Keywords*:** Polydopamine, mussel-inspired, adhesion, biomedicine, dentistry

4.1 Introduction

In the recent years, researchers across the globe are focussing on biopolymers as well as bioinspired polymer materials for different applications. After the cellulose, chitosan is the second largest biopolymer, which is used widely for biomedical applications. Chitin is found in the exoskeleton of crustaceans, which on deacetylations yields chitosan. Chitosan is one of the important biopolymers, which is widely used as an adsorbent for water purification [1], for membrane applications [2–5], biomedical applications, etc. [6]. However, in view of its low mechanical strength, several chemical modifications were performed for improving its strength [7–9]. We also reported other bioinspired material, such as tannic acid [10].

Biological structures have changed dramatically over millennia. Nature has been proactive in fine-tuning material characteristics, resulting in the structure-function relationship being optimized. In keeping with this trend, mussels' capacity to cling to diverse moist surfaces with adequate strength to resist powerful ocean currents has spurred curiosity and research into the role of polydopamine (PDA). The extensively replicated 3,4-dihydroxy-L-phenylalanine (DOPA) motif in the sticky foot protein Mefp-5 (mytilusedulis foot protein 5) generated by mussels is responsible for these adhesive characteristics [11, 12]. Polydopamine was initially recognized to have an important role in sessile mussel adhesion, and it has received a lot of attention in the last decade [13]. It was initially examined for possibilities in nanotechnology after being used in coatings for bulk materials.

Dopamine is a neurotransmitter, which is widely utilized for modification of biomaterial surfaces, inspired by natural mussel adhesion phenomena. The polydopamine coating has been shown to operate on a variety of surfaces, including superhydrophobic materials for converting them into hydrophilic. As a result, a new approach has emerged for surface alteration, which has attracted a lot of attention, especially in the domains of materials science, biology, and biomedicine. Oxidative polymerization of dopamine forms polydopamine when exposed to aqueous solutions. This mimics

DOPA and has a high adhesive ability to different substrates under wet circumstances via covalent and noncovalent bonding [14, 15].

Surface modification based on polymerized dopamine has recently emerged as a simple, straightforward, safe, successful, and cost-effective solution for bone tissue engineering. The production of PDA coating on the biomaterial interface during polydopamine-assisted surface modification is highly influenced by dopamine oxidation in weakly alkaline solutions. Polydopamine coatings have a wide range of chemical capabilities, ranging from protection against negative effects of light to serve as biosensors that communicate biological data, drug targeting, tumor ablation, magnetic resonance imaging, CT imaging, etc.

Polydopamine-derived surfaces allow effective bioconjugation of various biomaterials for emerging applications, such as sterilization of surgical instruments, drug delivery systems, contrast agents, bioadhesives, and anti-protein-adsorption interfaces. In the dentistry field, mussel-inspired wet-adhesion has gained a stronghold in dental research as surface tissue adhesives and biomaterial surface modifiers. The wet bioadhesion capacity, Matrix-metalloproteinases (MMP) inhibitory activity, and biomimetic-remineralization ability make PDA a useful additive for dental adhesives [16, 17].

Also, coating polydopamine on dentin has induced remineralization, which may be employed as an effective remedial measure for dentin super sensitivity. This review article emphasizes on latest advances in PDA applications, further providing clarity on its structure and properties. Apart from material make-up, other constituents of PDA properties are condensed—notably nanostructure, availability, mechanics, and how they could potentially be emerged as a tool for many biomedical and dental needs.

4.2 Structure of PDA

Literature suggests the mussel adhesive protein can generate adhesive "feet" to connect to various objects in damp and even rigorously trembling environments. Polydopamine's adhesive properties make it a valuable tool as a biosubstitute surface transformer and biomolecule immobilizer. However, because of the difficulties posed in the structural study of polydopamine and similar molecules, such as eumelanin, the binding pathway of polydopamine on the conjunction of different substrate platforms is again under debate [15, 18, 19]. The comparison of polydopamine to eumelanin is a good way to understand how a rich layer of polydopamine is formed chemically. L-3,4-dihydroxyphenylalanine (LDOAP) or tyrosine are used as an antecedent in the spontaneous

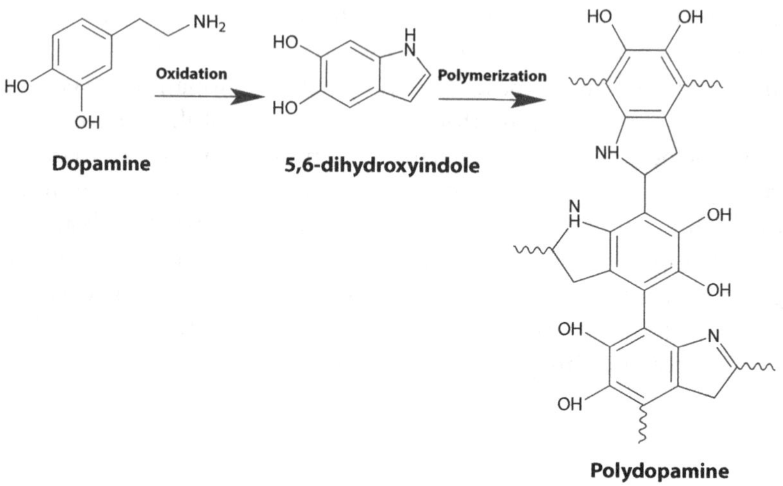

Figure 4.1 Oxidative polymerization of dopamine to polydopamine. (Journal-Frontiers in bioengineering and biotechnology. Publisher- Frontiers Media SA.)

formation of eumelanin. These DA-based monomers, as well as other essential PDA polymerization conditions, constitute critical elements in the synthesis of PDA [20]. In an aqueous solution, under the conditions of oxidation, self-polymerization of dopamine yields polydopamine (PDA) (Figure 4.1) [12].

4.3 Polydopamine as a Biomedical Material

PDA is a green, environment-friendly substance, accessible and very simple. It offers biocompatibility, which is non-toxic to living tissues. It enhances the hydrophilicity of surfaces upon deposition. The intrinsic antimicrobial property of PDA enables it for application in bioscience [21, 22]. PDA and the derivatives have simple and varied processing conditions, as well as high adhesion and hence they are appealing coatings for surface functionalization. It also includes functional groups, which may be utilized in the modification of surfaces. PDA along with the derivatives have a potential that is distinct as a material that aids in the functionalization of medical devices due to their strong adhesion to a variety of polymers and commodity metals, precursors found in nature (e.g., norepinephrine, L-dopa, dopamine), and the bioconjugation flexibility of hydroxyl groups and primary amines. The evolution of polydopamine as a covering layer for medical materials is motivated by its unique combination of benign and simple manufacturing, widespread adherence to various commodity materials, and varied functional groups [23].

4.4 Polydopamine as a Biomedical Adhesive

The use of a PDA as a functional surface was one of the first applications of the device. A high molar concentration of sticky catechols, as well as strong intermolecular bonding, texture, and mechanics, contribute to polydopamine's adhesive qualities. Because the elastic modulus of PDA can exceed E_{PDA} ~2 GPa in both real and theoretical estimates, binding in PDA derived coatings is influenced by the roughness [24, 25].

Polymer pillars coated with PDA produced by Lee *et al.* were employed as a biological adhesive. It was shown that this coating improved the adhesive properties (substrate to substrate) by a factor of 15 [12]. The interaction's durability was enhanced greatly in addition to its strength. PDA deposition on nearly any surface was described by Lee *et al.* by simple placement in a dopamine-based buffered solution. Then it was revealed that this treatment boosted the chances of grafting [12]. For example, enzymes, polysaccharides, various polymers, DNA, proteins, and nanoparticles have all been implanted onto a PDA covering [12, 26–30]. Ma *et al.*, for example, employed PDA's adhesive characteristics to improve the binding effectiveness of environmental contaminants to agglomerates of superparamagnetic iron oxide nanoparticles [31].

The use of polydopamine as a component in hydrogel composites is another example. Cross-linking with Fe^{3+} ions occurs when catechol groups are present, which is comparable to the function of the mussel byssus. Lu and colleagues added a layer of PDA on clay to further illustrate this. The clay particles that have been changed were subsequently combined with epoxy resin, resulting in increased particle dispersion and interfacial stress transmission, as well as improved overall material properties [32]. By covering titanium oxide sheets with polydopamine, Nam *et al.* enhanced dye-sensitized solar cells. The power conversion efficiency of the coated dye-sensitized solar cells was determined to be 1.2% [33]. It has also been studied as a solar fluid dopant for solar thermal energy generation as a photovoltaic substitute.

The efficiency of PDA nanoparticles, which are made up of PDA and bovine serum albumin, was investigated using the solar simulator as an irradiation source in a 3D-print-based flow circuit. When NPs were added to the solar fluid, it increased its efficiency by up to 280% over ordinary solar-based fluid [34].

Nano-sized PDA/BSA (sPDA/BSA) nanoparticles increased the solar fluid's heating performance when compared to silver nanoparticles or simple solar-based liquid made up of water and glycol. The research also confirmed that the material composed of PDA NPs, determined specific heating rate was size-dependent. PDA was examined as a stabilizer for polyethylene glycol and dextran inclusive aqueous solutions in a study by Zhang *et al.* [35]. Following surfactant addition, the synthesized capsules remained intact, droplets swelled to a little degree were seen after dilution with water, demonstrating the emulsion's increased stability.

PDA's inherent mechanical qualities and roughness at interface restrict its usability as a frequently used adhesive, despite its high density of sticky catechol motifs [24, 25]. PDA's unique qualities have aided in the advancement of numerous technologies. Additionally, due to features such as biocompatibility, biodegradability, and adaptability, PDA is being studied and used in a variety of biomedical applications. This part has been explained in detail in the next section.

4.5 Availability of Polydopamine and Its Biomedical Applications

Polydopamine is available as coatings of nanomaterials, capsules, nanoparticles, and nanocomposites. PDA's physicochemical features, which include high adherence to various surfaces, the capacity to undergo post-polymerization synthetic alteration, and the existence of stable radicals in the

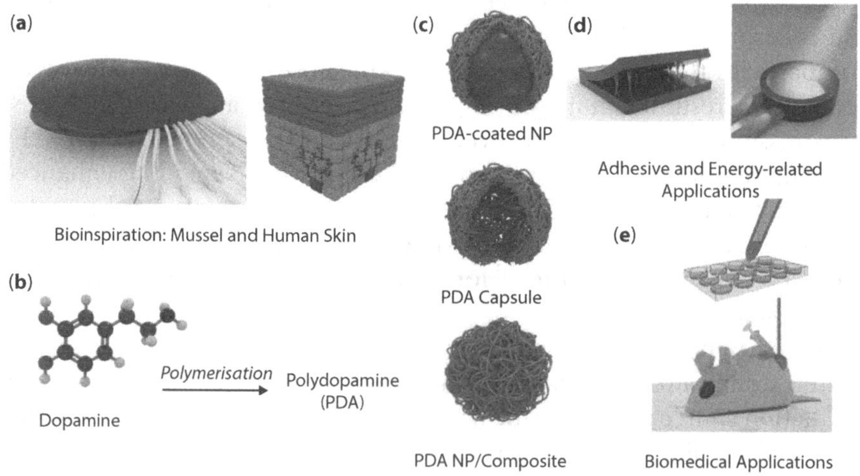

Figure 4.2 Mussel and human skin-boinspiration (b). Self-polymerization of dopamine to polydopamine (c) Depiction of generation of varying types of structures of PDA namely – NP coated with PDA, capsule of PDA, and composite of PDA NP, (d&e) Different PDA structures as robust tools for adhesive and in energy-derived and biomedical needs [38]. (Journal-Materials (Basel). Publisher-MDPI.)

polymer structure, have led to a wide range of proposed and demonstrated uses [36–38] (Figure 4.2).

4.6 Polydopamine Coatings of Nanomaterials

Polydopamine has widely found acceptance to enhance and/or alter the characteristics of nanomaterials. The coating procedure is simple to follow and may be used on any surface. It suffices to immerse such nanomaterials in a dopamine-based solution to coat them with PDA. Materials submerged in dopamine solution at a 2 mg per millilitre concentration of 10 Mm tris(hydroxymethyl) aminomethane (TRIS), optimization of pH to 8.5 resulted in a PDA layer deposition of 50 nm thickness, according to Messersmith *et al.* [12]. For the production of PDA films, these reaction conditions are still the most common [39]. Furthermore, they may be used for both bulk materials and nanoparticles. Spin-coating PDA onto surfaces is another option; however, this method is mostly employed for electrical applications, thus its utility in the biomedical sector is limited.

It has proven to be nonreactive and biocompatible to tissues when applied as a coating [40]. Ku *et al.* (2010) established the biocompatibility of PDA *in vitro* on diverse cell types (e.g., neurons, fibroblasts, osteoblasts, and endothelial cells) and observed excellent adherence of PDA coated material onto these cells with no adverse effects [41].

Bacterial adherence is increased when coatings are coated with PDA. Contact-mediated antifungal activity was observed when a single bacterium was immobilized with a 10-fold on a PDA-coated surface. PDA coatings have been proven to be effective in a variety of biosensing applications [16, 42–44]. The requirement to enhance and generate more sensitive and selective high throughput biosensing tools for diagnostics, drug optimization, and drug discovery continues to be critical. PDA coatings are the final piece of the biosensor puzzle, and they have been utilized in a variety of ways, including as a matrix for molecular imprinting, as a coating in an amperometric biosensor, as a new label, and to encapsulate entire cells. It is used quite extensively for targeting drug delivery, primarily due to its optimal characteristics (i.e., biodegradation, biocompatibility, and versatility), and this desirable feature shows promise for biomedical applications of nanoparticles with the eventual goal of pin pointedly targeting cancer cells and attenuating side effects [45, 46]. In addition, several studies have demonstrated that PDA may be utilized well for magnetic resonance imaging [47]. Various nanoparticles, such as gold, iron oxide NPs, Prussian blue NPs, silica NPs, MnO_2 NPs coated with PDA, display photothermal and photodynamic therapy, CT imaging and tumor ablation characteristics and magnetic resonance imaging demonstrate chemotherapeutic activities by suppressing tumor development [48, 49]. Complex gold nanoparticles with a homogenous PDA coating is also termed gold nanostars, have great optical characteristics and are often used for theranostic-based applications [50].

4.7 Polydopamine-Based Capsules

Other formulations in the form of PDA capsules have been synthesized in addition to the coatings [51]. Polymer capsule synthesis is often complicated, labor-intensive, and hence costly [52]. On the other hand, the production of PDA capsules is rather uncomplicated [53]. Furthermore, the benefits of producing capsules from PDA are enhanced by its unique characteristics—namely drug delivery and the concurrent potential of ligand grafting. Templates are used to make the majority of PDA capsules. In general, PDA is deposited onto a template nanostructure, which is then removed using optimized techniques. So far, many variations of this method have been described. Calcium carbonate microparticles, magnesium carbonate particles, silica nanoparticles, and polystyrene spheres, for example, are utilized as a template to create PDA capsules catering to various sizes and shapes. Additionally, these PDA capsules have shown to have no substantial influence on cell viability *in vitro* as suggested by Postma *et al.* in 2009 [54]. For drug delivery, PDA-derived capsules were employed or coupled with functional drugs. *In vitro*, capsules with doxorubicin have been shown to neutralize HeLa cancer-affected cells more effectively than free doxorubicin of the same dosage killing tumor tissue.

4.8 Polydopamine Nanoparticles and Nanocomposites

PDA has the unique capability of efficiently converting light into heat. PDA nanoparticles (NP) can be used to provide an effective new platform for photothermal treatment. The nanoparticles can further be utilized for biological applications—medication administration, imaging due to additional advantageous characteristics (namely biocompatibility). Different techniques for producing PDA NPs are discussed based on the intended application size. Dopamine polymerizes in a combination of ethanol and water solution with minimal reaction modifications is a frequently utilized technique (e.g., pH and reaction duration) [55, 56]. This reaction yields NPs varying in size from 70 to 400 nm, with the size of the NPs determined by ethanol by water ratios and additionally by dopamine concentration [57]. A combination of protein solutions and dopamine in TRIS-buffered water at a pH of 8.5 is another option. The ratio of dopamine to protein concentrations may be used to adjust the size, with the additional benefit of integrating functional proteins into PDA-based nanoparticles. PDA NPs, like their coatings and capsules, were shown to have no negative impact on cell viability when they were internalized [58]. Zhang *et al.* 2012 treated mouse cells to escalating doses of polydopamine (10–160 μg/mL), and no toxicity was detected [59]. PDA NPs are used for tumor ablation targeting cancer cells resulting in cell death which could be due to lysosomal membrane permeabilization [60]. It has been successful in drug delivery and has shown to be very effective to deliver the loaded drugs to target tissues [61]. Using photothermal therapy and magnetic resonance imaging, PDA NPs for tumor growth *in vivo* were entirely overcome [62]. Elimination of bacterial biofilm by producing ROS and causing destruction of the membrane [63]. Nanocomposites have shown a positive response in cases of Alzheimer's disease by suppressing the Aβ production and aggregation. So it is suggested as a pharmacological modality in the treatment of Alzheimer's disease [64].

4.9 Polydopamine Properties

4.9.1 Cell Adhesion

The coating of PDA is an effective approach for changing a range of bioinert substrates into bioactive substrates, including some surfaces that do not get moist and 3D porous derived scaffolds, by boosting cell binding of diverse cell structures—namely osteoblasts, chondrocytes, and pheochromocytoma. On PDA-coated surfaces, endothelial cells and fibroblast cells showed significantly improved adherence and vitality [65]. Additionally, PDA coatings have found usage for cell patterning using a variety of methods, including photolithography, microfluidics, and micro-contact printing, because of their great affinity for cells as well as their excellent stability and material independence [65].

Aside from cell patterning, the well-arranged PDA sheath can be utilized for immobilization of protein, thiol or amine conjugation, and metal nanoparticle immobilization to construct a variety of chemical patterns on substrates [66].

The combined effects of PDA modification on endothelial cell (EC) and smooth muscle cell (SMC) function on titanium dioxide nanotubes without changing the nanostructure morphology was examined by Zhong *et al.* [67]. They discovered that PDA modification and nanostructures worked together to increase EC binding, proliferation, migration, and discharge of nitric oxide. Meanwhile, SMC adherence and proliferation were inhibited by the PDA-modified nanotube surface [67].

The putative mechanism for PDA coatings enhancing cell adherence has also been explored. Because serum protein adsorption happening on both unaltered and PDA-coated substrates, Ku *et al.* speculated that the surface modification through PDA arrested probable protein denaturation, which is thought to be the cause of improved cell adhesion [65, 68].

Tsai and group members evaluated cell adherence on a PDA-coated surface when serum was present versus when it was not [68]. They discovered that the presence of serum proteins significantly increased cell adherence to PDA-coated surfaces compared to the non-appearance of serum proteins.

Furthermore, the adsorption of fibronectin was greater on PDA-coated sheaths than on unaltered surfaces. It suggests that augmenting the immobilization and/or adsorption of adhesive proteins, such as fibronectin, onto substrates could potentially increase cell adherence to PDA-coated surfaces.

4.9.2 Mineralization and Bone Regeneration

PDA coatings contain many catecholamine moieties, which are responsible for the functionalization of chemicals of the wide variety of material surfaces, as well as the formation of hydroxyapatite when calcium ions are concentrated at the interface layer. PDA coated titanium cultivated in simulated physiological fluid for 2 weeks was homogeneously covered with calcium phosphate salts in research by Ryu *et al.*, compared to the restricted mineral deposition on the pure titanium substrate (Figure 4.3) [69].

One of the intriguing works was using PDA to connect the mineral sand polymeric fibers, demonstrating biomineralization of electrospun fibers. The mineral coatings increase the ultimate tensile strength, stiffness, and toughness of polymeric fibers, making them more similar to the mechanical characteristics of real bone. PDA coatings exhibited favorable adhesion and osteogenic cell formation of various cell types *in vitro*, in addition to promoting mineralization [70, 71].

According to Wu *et al.*, porous silicon dioxide scaffold coated with polydopamine demonstrated an uptick in adhesion, ossification of human bone marrow stromal cells (hBMSCs), proliferation—showcasing improved osteogenic property [71].

Rim *et al.* also used PDA coatings on eletrospun fibers to distinguish the ossification of the human mesenchymal stem cells. Consequently, they discovered adding coatings of PDA to the cells increased cell attachment,

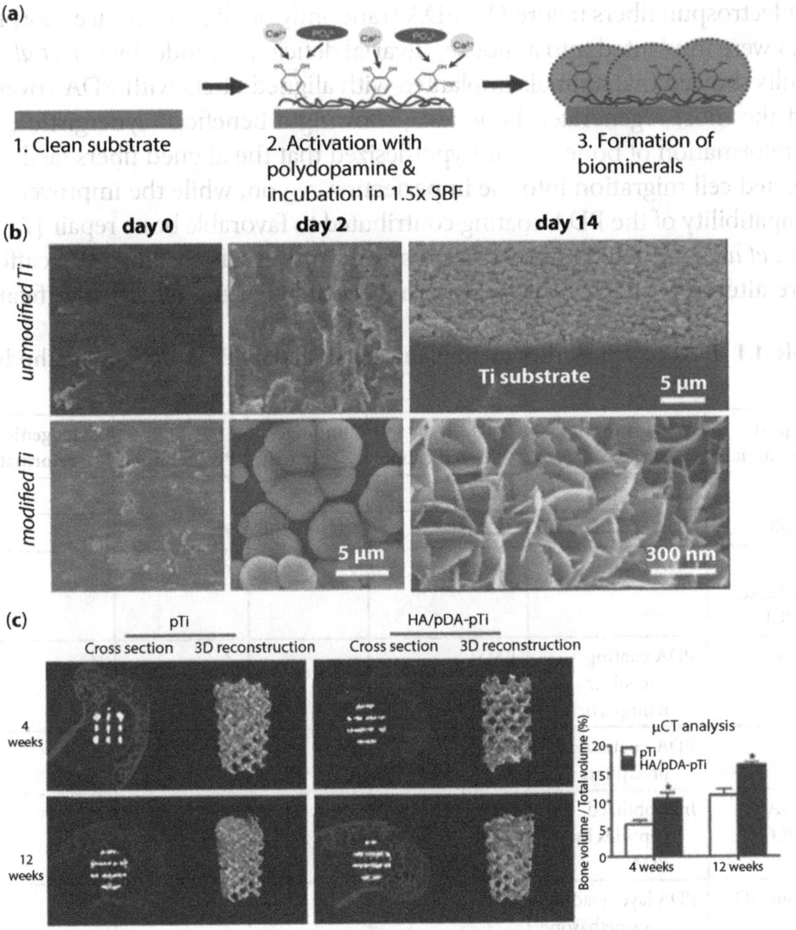

Figure 4.3 The calcification and reformation of bone rejuvenated by the modification using PDA coating. (a) Pathway for calcium phosphate (CaP) crystal generation by the PDA-aid. (b) FESEM pictures depicting the effects of the surface modification through PDA on the generation of CaP minerals after incubation the substrates in simulated body fluid (SBF). (c) Micro-CT images were captured of porous Ti6Al4V scaffolds (pTi) and HA/PDA-pTi scaffolds after 4 and 12 weeks of implantation. The yellow color in the images indicates newly formed bone. The regenerated bone volume/total volume percentages were quantified for these implants. Statistical significance was observed (* indicates) when compared to the pTi group, with a significance level of $p < 0.05$. Taken from Ref. [69]. Reference article [72]. (Journal- Biosurface and Biotribology. Publisher-Science Direct.)

proliferation—it also improved osteogenic differentiation with increasing ALP activity and osteogenic gene expression [73]. Distinct types of bone-derived cells, such as MG-63 human osteoblastic cells and mice osteoblastic cells, have also been shown to be favorably stimulated by PDA coatings [74, 75].

Electrospun fibers tailored by PDA (randomly or aligned surface morphology) were implanted into a mouse calvarial deficiency model by Lee *et al.* The results showed that animals implanted with aligned fibers with PDA covering had the most regenerated bone area, showing a beneficial synergistic effect on reformation of bone. It was hypothesized that the aligned fibers facilitated directed cell migration into the imperfection region, while the improved biocompatibility of the PDA coating contributed to favorable bone repair [76].

Li *et al.* conducted another fascinating work in which Ti6Al4V scaffolds were altered by PDA-directed hydroxyapatite coating, which significantly

Table 4.1 *In vitro* cell studies on implants decorated by PDA-derived methods [78].

Support materials	What is in the surface	Cell specification	Binding capacity	Proliferation	Osteogenic reformation
PCLDA	PDA coating	hMSCs	✓		✓
Parafilm	PDA coating	ADMSCs	✓	✓	✓
Calcium silicate/ PCL	PDA coating	WJMSCs	✓	✓	✓
Alginate	PDA coating and silver nanoparticles	BMSCs		✓	✓
PCL	PDA-coated/HA precipitate	hMSCs	✓	✓	✓
op-HA/ PLGA	Immobilized peptides by PDA	MC3T3-E1	✓	✓	✓
Porous SiO_2	PDA layer loading dexamethasone	BMSCs	✓	✓	✓
B-TCP	PDA NPs	BMSCs	✓	✓	✓
Ti	PDA/HA coatings	MG63	✓	✓	✓
Ti	PDA/BMP-2 coating	PDLSCs			✓
Ti	Microporous PDA architectures adsorbing BMP-2	BMSCs	✓	✓	✓
Ti	PDA/HA/BMP-2	BMSCs	✓	✓	✓

NPs, nanoparticles; PDA, polydopamine; HA, hydroxyapatite; BMP-2, bone morphogenic protein-2; BMSCs, bone marrow mesenchymal stem cells; PCLDA, poly (£-caprolactone diacrylate; PCL, polycaprolactone; β-TCP, β-tricalcium phosphate; hMSCs, human mesenchymal stem cells; ADMSCs, adipose-derived mesenchymal stem cells; PDLSC, periodontal ligament stem cell; PLGA, poly(lactic-co-glycolic acid); WJMSC, Wharton's jelly mesenchymal stem cells Journal- Journal of orthopedic translation. Publisher-Elsevier.

improved osteointegration and encouraged bone restoration after implantation into rabbit femoral condylar imperfections at 4 and 12 weeks [77].

Table 4.1 shows wide variety of implant materials decorated by PDA-derived surface modifications and their applicability [78].

4.9.3 Blood Compatibility

In a study conducted by Ku and Park in 2010, PDA coatings were used to help polycaprolactone nanofiber scaffolds re-endothelialize. PDA coatings improved endothelial cell (EC) adherence, spreading, vitality, and positive expression of EC markers as compared to virgin polycaprolactone scaffolds, suggesting that they might be used in vascular tissue engineering. Critically, the latest research has shown that PDA coatings apart from improving EC development and function, also decrease smooth muscle cell proliferation [41].

According to Yang *et al.*, coatings of PDA on stents of 316 L stainless steel dramatically improved endothelialization, concurrently limiting SMC proliferation and adhesion. The coatings of PDA on stents, in particular, demonstrated resistance and good stability to stent compression and expansion deformation [79].

According to Luo *et al.*, heat oxidation in an air changed the major catechol and amino bands in PDA coatings into quinone bands. Thermic oxidation of coatings of PDA boosted the potential to stimulate EC development while diminishing the repression of SMC proliferation, according to the authors. They concluded that phenolic, quinone groups are directly linked to the opposing effects of coatings of PDA on SMCs and ECs [80].

Ding *et al.* researched correlations between "surface property—protein adsorption—vascular cell function" of polydopamine-based coatings produced with different starting concentrations of dopamine. The findings revealed that the quinone group on coatings of PDA caused more protein adsorption, which in turn aided EC adhesion and growth [81].

4.9.4 Antimicrobial Effect

PDA has an antimicrobial effect as well. According to one hypothesis, dopamine self-polymerization on the bacterium encapsulates the bacteria inside the PDA film, and the PDA shell layer inhibits the bacteria from reproducing. Another idea claimed that a layer of PDA established a membrane with poor permeability to certain components, limiting nutrients from being transported and metabolic waste from being expelled.

Coatings of PDA on gold/polypyrrole bi-layer actuators have been shown to enhance the efficacy of capturing *Escherichia coli* (*E. coli*) bacteria from physiological medium while preserving actuation performance due to their high bioactivity [42].

When compared to a pure zirconia surface, Liu *et al.* discovered that PDA-coated zirconia substantially lowered *Streptococcus gordonii* and *Streptococcus* mutans growth, despite the PDA coating not influencing bacterial survival [82].

Rough PDA (rPDA) was employed in a recent study to demonstrate the antibacterial characteristics of PDA inclusive coatings made with a simple shaking-based procedure. As part of this study, a shaking solution condition was used instead of a static solution contained condition in the standard synthesis method for smooth PDA-based coatings (sPDA), resulting in rPDA coatings. The rPDA coatings possessed identical chemical properties as the smooth PDA-based coatings, however, there was the formation of radically differing morphology and wettability, according to Raman spectra analysis. Surprisingly, rPDA coatings demonstrated much stronger antibacterial activity against gram-negative *E. coli*, gram-positive *Staphylococcus aureus,* and *Pseudomonas aeruginosa* than smooth PDA-based coatings and unaltered surfaces. The contact killing, according to the authors, was caused by damage to the cell membranes of bacteria cultured with the rPDA coatings. The physical ties of PDA were found to be important in this study [83].

Additionally, the antibacterial efficiency of PDA-based coatings on the material's contact surface has not been adequately evaluated, and the results are inconsistent, necessitating more research in this area.

4.10 Dental Applications

The dental applications for polydopamine have not been explored entirely. Available data on its uses are mostly based on the mussel-adhesion phenomenon of polydopamine in a wet environment. Recent research has been conducted to improve the dentin adhesive interface coupled with significant results of improved strength and durability. Another application that has been researched is its use in enamel and dentin remineralization.

4.11 Dental Adhesives

The hybrid layer (HL) was formed by the efficient and stable impregnation of resin monomers into the dentin substrate, which couples adhesive to the underlying mineralized dentin. The structural integrity, mechanical stability, and enzymatic resistance of the HL all have a role in the resin-dentin bond interface's durability. To chemically dehydrate and decrease the hydrophilicity of the demineralized collagen matrix, ethanol-wet-bonding was added [84]. It produced larger inter-fibrillar gaps, allowing hydrophobic resins to completely permeate the matrix. According to molecular dynamic simulations, ethanol replaces just the surface layer of firmly bound water without altering the interior two layers of bound water [85]. As a result, an adhesive that is resistant to both hydrolytic and proteolytic degradation in a wet oral environment and has biomimetic remineralization capacity can provide long-term bond durability. PDA is a synthetic analog of mussel-adhesive-foot-proteins (Mefps), which are produced by oxidative polymerization of dopamine (DOPA) and have a high adhesion capability to any sort of surface when wet. This sticky protein's catechol groups dispense securely bound water molecules and establish hydrogen bonds with substrates. By overcoming repulsive hydration forces, displacing surface-bound water, and creating strong surface adhesion, increased bioadhesion is achieved [86]. Due to the matrix metalloproteinase (MMP) inhibitory effect, PDA integrated self-etch and total-etch adhesives can serve as a possible addition to increase bond strength and durability without affecting the degree of polymerization and by increasing collagen stability. PDA is a helpful component for dental adhesives because of its wet bioadhesion capability, MMP inhibitory action, and biomimetic-remineralization ability [87].

4.11.1 Tooth Mineralization

In recent decades, the re-mineralization of superficial dental tissue has gained significant attention as one of the non-invasive methods, and its therapeutic value has been widely acknowledged. The polydopamine covering enhances remineralization of demineralized dentin by occluding dentin tubules with densely packed HA crystals.

The tubules in the polydopamine-coated dentin samples were clogged by definite crystals after seven days of remineralization., and though less crystals were seen on enamel it was abundantly seen on dentin (Figure 4.4a, c, and d). Furthermore, in the specimens without polydopamine covering, minor crystal precipitates were detected on the dentin surface as well as its tubules (Figure 4.4b).

Catecholamines at the interfaces are implicated in the binds because dopamine self-polymerizes *in situ*. Surface-anchored catecholamine moieties (catecholamines that are not involved in substrate adhesion) adhere to Ca^{2+} and enhance the interface with calcium ions, allowing hydrxyapatite (HA) crystals to form more easily [69]. On enamel, there is not much of a difference. Hydroxyapatite crystals originate from its existing nucleus

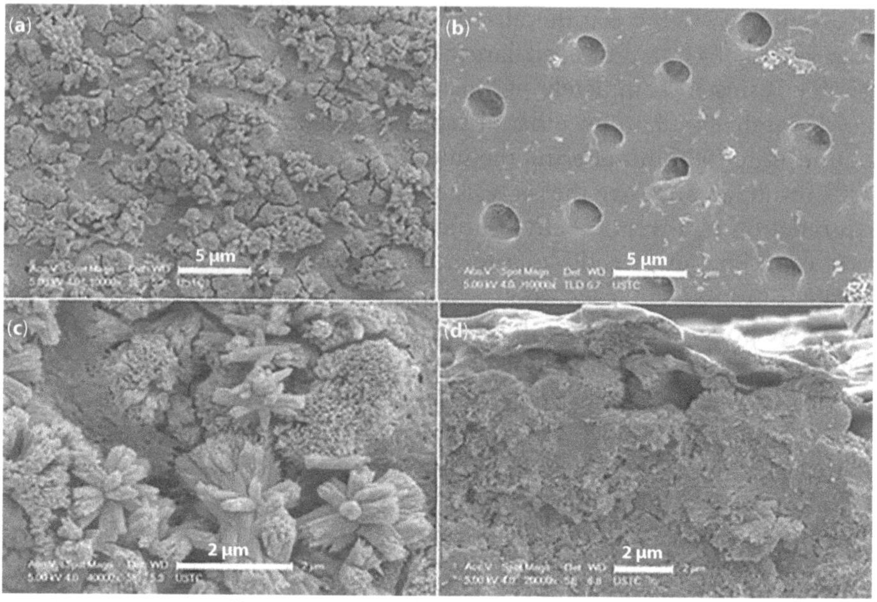

Figure 4.4 FESEM pictures of dentin surface calcification after remineralization for seven days with the polydopamine assisted surface coating (a, c, d), without coating (b) as presented in [87]. (Journal- ACS Appl Mater Interfaces. Publisher-American Chemical Society.)

and they grow perpendicular to enamel surface and fill up tightly with each other during enamel remineralization. During dentin remineralization, HA crystals formed more specifically in the tubules of dentin rather than in the peritubular dentin and in a distinct manner. As a result, dentin remineralization differs from enamel remineralization in terms of mechanism [88–90].

4.12 Conclusions

PDA is an outstanding example of bioinspiration and effective translation of natural phenomena into usage in biomedical applications. It is extremely malleable and provides a promising platform in the form of the nanocomposite, nanoparticle, or as a coating of existing materials. Dopamine's adhesive capabilities, as shown in mussels, have inspired its usage as molecular glue, therefore adding new or increasing the coated material's existing characteristics. The additional PDA properties in biocompatibility, biodegradability, anti-microbial activity, bone regeneration, and versatility make it a promising material that can be mapped into various fields of biomedicine. Although the area of PDA-based nanomaterials is in its infancy, current examples and applications have demonstrated its promise. The studies on the use of PDA in dentistry are limited to dental adhesives and remineralization to date. The versatile properties demonstrated can further be researched and explored for different applications in the field of dentistry.

References

1. Kumar, R. *et al.*, Polysulfone–chitosan blend ultrafiltration membranes: Preparation, characterization, permeation and antifouling properties. *RSC Adv.*, 3, 21, 7855–7861, 2013.
2. Kumar, R. *et al.*, Permeation, antifouling and desalination performance of TiO2 nanotube incorporated PSf/CS blend membranes. *Desalination*, 316, 76–84, 2013.
3. Kumar, R. *et al.*, Synthesis and characterization of novel water soluble derivative of Chitosan as an additive for polysulfone ultrafiltration membrane. *J. Membr. Sci.*, 440, 140–147, 2013.
4. Kumar, R. *et al.*, Performance improvement of polysulfone ultrafiltration membrane using N-succinyl chitosan as additive. *Desalination*, 318, 1–8, 2013.
5. Shenvi, S.S. *et al.*, Preparation and characterization of PPEES/chitosan composite nanofiltration membrane. *Desalination*, 315, 135–141, 2013.
6. Zhao, D. *et al.*, Biomedical applications of chitosan and its derivative nanoparticles. *Polymers*, 10, 4, 462, 2018.
7. Padaki, M. *et al.*, Preparation and characterization of sulfonated polysulfone and N-phthloyl chitosan blend composite cation-exchange membrane for desalination. *Desalination*, 298, 42–48, 2012.
8. Padaki, M., Isloor, A.M., Wanichapichart, P., Polysulfone/N-phthaloylchitosan novel composite membranes for salt rejection application. *Desalination*, 279, 1, 409–414, 2011.
9. Kolangare, I.M. *et al.*, Antibiofouling hollow-fiber membranes for dye rejection by embedding chitosan and silver-loaded chitosan nanoparticles. *Environ. Chem. Lett.*, 17, 1, 581–587, 2019.
10. Ibrahim, G.P.S. *et al.*, Bio-inspired, fouling resistant, tannic acid functionalized halloysite nanotube reinforced polysulfone loose nanofiltration hollow fiber membranes for efficient dye and salt separation. *J. Water Process Eng.*, 20, 138–148, 2017.
11. Silverman, H.G. and Roberto, F.F., Understanding marine mussel adhesion. *Mar. Biotechnol.*, 9, 6, 661–681, 2007.
12. Lee, H. *et al.*, Mussel-inspired surface chemistry for multifunctional coatings. *Science*, 318, 5849, 426–30, 2007.

13. Liu, Y., Ai, K., Lu, L., Polydopamine and its derivative materials: Synthesis and promising applications in energy, environmental, and biomedical fields. *Chem. Rev.*, 114, 9, 5057–5115, 2014.
14. Cha, H.J., Hwang, D.S., Lim, S., Development of bioadhesives from marine mussels. *Biotechnol. J.*, 3, 5, 631–8, 2008.
15. Bernsmann, F. *et al.*, Characterization of dopamine–melanin growth on silicon oxide. *J. Phys. Chem. C*, 113, 19, 8234–8242, 2009.
16. Lynge, M.E. *et al.*, Polydopamine—A nature-inspired polymer coating for biomedical science. *Nanoscale*, 3, 12, 4916–4928, 2011.
17. Hebbar, R. *et al.*, Fabrication of antifouling, antimicrobial, well dispersed polydopamine functionalized halloysite nanotube-polyetherimide mixed matrix membranes for the heavy metal removal application. *J. Mater. Chem. A*, 4, 764–774, 2015.
18. Clark, M.B. *et al.*, Solid-state analysis of eumelanin biopolymers by electron spectroscopy for chemical analysis. *Anal. Chem.*, 62, 9, 949–956, 1990.
19. Dreyer, D.R. *et al.*, Elucidating the Structure of Poly(dopamine). *Langmuir*, 28, 15, 6428–6435, 2012.
20. d'Ischia, M. *et al.*, Polydopamine and eumelanin: From structure–property relationships to a unified tailoring strategy. *Acc. Chem. Res.*, 47, 12, 3541–3550, 2014.
21. Sahiner, N. *et al.*, Polydopamine particles as nontoxic, blood compatible, antioxidant and drug delivery materials. *Colloids Surf. B Biointerfaces*, 172, 618–626, 2018.
22. Huang, S. *et al.*, Polydopamine-assisted surface modification for bone biosubstitutes. *BioMed. Res. Int.*, 2016, 2389895, 2016.
23. Xiang, H. *et al.*, A redox cycling-amplified electrochemical immunosensor for α-fetoprotein sensitive detection via polydopamine nanolabels. *Nanoscale*, 10, 28, 13572–13580, 2018.
24. Chen, C.T. *et al.*, Self-assembly of tetramers of 5,6-dihydroxyindole explains the primary physical properties of eumelanin: Experiment, simulation, and design. *ACS Nano*, 7, 2, 1524–32, 2013.
25. Kwon, I.S. and Bettinger, C.J., Polydopamine nanostructures as biomaterials for medical applications. *J. Mater. Chem. B*, 6, 43, 6895–6903, 2018.
26. Lee, H., Rho, J., Messersmith, P.B., Facile conjugation of biomolecules onto surfaces via mussel adhesive protein inspired coatings. *Adv. Mater.*, 21, 4, 431–434, 2009.
27. Shin, Y.M. *et al.*, Bio-inspired immobilization of cell-adhesive ligands on electrospun nanofibrous patches for cell delivery. *Macromol. Mater. Eng.*, 298, 5, 555–564, 2013.
28. Ham, H.O. *et al.*, Facile DNA immobilization on surfaces through a catecholamine polymer. *Angew. Chem. Int. Ed. Engl.*, 50, 3, 732–6, 2011.
29. Wu, J. *et al.*, Mussel-inspired chemistry for robust and surface-modifiable multilayer films. *Langmuir*, 27, 22, 13684–13691, 2011.

30. Long, Y. *et al.*, Rapid sintering of silver nanoparticles in an electrolyte solution at room temperature and its application to fabricate conductive silver films using polydopamine as adhesive layers. *J. Mater. Chem.*, 21, 13, 4875–4881, 2011.
31. Ma, Y.-r. *et al.*, Polydopamine-coated magnetic nanoparticles for enrichment and direct detection of small molecule pollutants coupled with MALDI-TOF-MS. *ACS Appl. Mater. Interfaces*, 5, 3, 1024–1030, 2013.
32. Yang, L. *et al.*, A biomimetic approach to enhancing interfacial interactions: Polydopamine-coated clay as reinforcement for epoxy resin. *ACS Appl. Mater. Interfaces*, 3, 8, 3026–32, 2011.
33. Nam, H.J. *et al.*, A new mussel-inspired polydopamine sensitizer for dye-sensitized solar cells: Controlled synthesis and charge transfer. *Chem. - Eur. J.*, 18, 44, 14000–14007, 2012.
34. Hauser, D. *et al.*, Polydopamine nanoparticle doped nanofluid for solar thermal energy collector efficiency increase. *Adv. Sustainable Syst.*, 4, 1, 1900101, 2020.
35. Zhang, J. *et al.*, Water-in-water pickering emulsion stabilized by polydopamine particles and cross-linking. *Biomacromolecules*, 20, 1, 204–211, 2019.
36. Dreyer, D.R. *et al.*, Perspectives on poly(dopamine). *Chem. Sci.*, 4, 10, 3796–3802, 2013.
37. Lynge, M.E. *et al.*, Polydopamine/liposome coatings and their interaction with myoblast cells. *ACS Appl. Mater. Interfaces*, 3, 6, 2142–7, 2011.
38. Hauser, D. *et al.*, From bioinspired glue to medicine: Polydopamine as a biomedical material. *Mater. (Basel)*, 13, 7, 1730, 2020.
39. Ball, V. *et al.*, Kinetics of polydopamine film deposition as a function of pH and dopamine concentration: Insights in the polydopamine deposition mechanism. *J. Colloid Interface Sci.*, 386, 1, 366–72, 2012.
40. Hong, S. *et al.*, Attenuation of the *in vivo* toxicity of biomaterials by polydopamine surface modification. *Nanomedicine (Lond)*, 6, 5, 793–801, 2011.
41. Ku, S.H. *et al.*, General functionalization route for cell adhesion on non-wetting surfaces. *Biomaterials*, 31, 9, 2535–41, 2010.
42. Liu, A. *et al.*, Polypyrrole actuator with a bioadhesive surface for accumulating bacteria from physiological media. *ACS Appl. Mater. Interfaces*, 1, 4, 951–5, 2009.
43. Kang, S. and Elimelech, M., Bioinspired single bacterial cell force spectroscopy. *Langmuir*, 25, 17, 9656–9659, 2009.
44. Paulo, C.S., Vidal, M., Ferreira, L.S., Antifungal nanoparticles and surfaces. *Biomacromolecules*, 11, 10, 2810–7, 2010.
45. Baselga, J. *et al.*, AACR cancer progress report 2015. *Clin. Cancer Res.*, 21, 19 Supplement, S1–S128, 2015.
46. Misra, R., Acharya, S., Sahoo, S.K., Cancer nanotechnology: Application of nanotechnology in cancer therapy. *Drug Discovery Today*, 15, 19–20, 842–50, 2010.

47. Zeng, Y. *et al.*, Lipid-AuNPs@PDA nanohybrid for MRI/CT imaging and photothermal therapy of hepatocellular carcinoma. *ACS Appl. Mater. Interfaces*, 6, 111, 2014.
48. Li, Y. *et al.*, Structure and function of iron-loaded synthetic melanin. *ACS Nano*, 10, 11, 10186–10194, 2016.
49. Liu, R. *et al.*, Core-shell Fe3O4 polydopamine nanoparticles serve multipurpose as drug carrier, catalyst support and carbon adsorbent. *ACS Appl. Mater. Interfaces*, 5, 18, 9167–71, 2013.
50. Li, D. *et al.*, Construction of polydopamine-coated gold nanostars for CT imaging and enhanced photothermal therapy of tumors: An innovative theranostic strategy. *J. Mater. Chem. B*, 4, 23, 4216–4226, 2016.
51. Park, J. *et al.*, Polydopamine-based simple and versatile surface modification of polymeric nano drug carriers. *ACS Nano*, 8, 4, 3347–3356, 2014.
52. Campbell, J. and Vikulina, A., Layer-by-layer assemblies of biopolymers: Build-up, mechanical stability and molecular dynamics. *Polymers*, 12, 1949, 2020.
53. Cheng, W. *et al.*, Versatile polydopamine platforms: Synthesis and promising applications for surface modification and advanced nanomedicine. *ACS Nano*, 13, 8, 8537–8565, 2019.
58. Wang, Y. and Caruso, F., Template synthesis of stimuli-responsive nanoporous polymer-based spheres via sequential assembly. *Chem. Mater.*, 18, 17, 4089–4100, 2006.
55. Ho, C.-C. and Ding, S.-J., The pH-controlled nanoparticles size of polydopamine for anti-cancer drug delivery. *J. Mater. Sci. Mater. Med.*, 24, 113, 2013.
56. Wang, X. *et al.*, Multi-responsive photothermal-chemotherapy with drug-loaded melanin-like nanoparticles for synergetic tumor ablation. *Biomaterials*, 81, 114–124, 2016.
57. Yan, J. *et al.*, Polydopamine spheres as active templates for convenient synthesis of various nanostructures. *Small*, 9, 4, 596–603, 2013.
58. Liu, Y. *et al.*, Dopamine-melanin colloidal nanospheres: An efficient near-infrared photothermal therapeutic agent for *in vivo* cancer therapy. *Adv. Mater.*, 25, 9, 1353–1359, 2013.
59. Zhang, X. *et al.*, Biocompatible polydopamine fluorescent organic nanoparticles: Facile preparation and cell imaging. *Nanoscale*, 4, 18, 5581–5584, 2012.
60. Hauser, D. *et al.*, Polydopamine/transferrin hybrid nanoparticles for targeted cell-killing. *Nanomaterials*, 8, 12, 1065, 113, 2018.
61. Li, Y. *et al.*, Targeted polydopamine nanoparticles enable photoacoustic imaging guided chemo-photothermal synergistic therapy of tumor. *Acta Biomater.*, 47, 124–134, 2017.
62. Liu, F. *et al.*, Controllable synthesis of polydopamine nanoparticles in microemulsions with pH-activatable properties for cancer detection and treatment. *J. Mater. Chem. B*, 3, 33, 6731–6739, 2015.

63. Yuan, Z. *et al.*, Near-infrared light-triggered nitric-oxide-enhanced photodynamic therapy and low-temperature photothermal therapy for biofilm elimination. *ACS Nano*, 14, 3, 3546–3562, 2020.
64. Srivastava, A.K., Roy Choudhury, S., Karmakar, S., Near-infrared responsive dopamine/melatonin-derived nanocomposites abrogating *in situ* amyloid β nucleation, propagation, and ameliorate neuronal functions. *ACS Appl. Mater. Interfaces*, 12, 5, 5658–5670, 2020.
65. Ku, S.H. and Park, C.B., Human endothelial cell growth on mussel-inspired nanofiber scaffold for vascular tissue engineering. *Biomaterials*, 31, 36, 9431–7, 2010.
66. Chien, H.-W. *et al.*, Tunable micropatterned substrates based on poly(dopamine) deposition via microcontact printing. *Langmuir*, 28, 13, 5775–5782, 2012.
67. Zhong, S. *et al.*, Effects of polydopamine functionalized titanium dioxide nanotubes on endothelial cell and smooth muscle cell. *Colloids Surf. B Biointerfaces*, 116, 553–60, 2014.
68. Tsai, W.B. *et al.*, Poly(dopamine) coating of scaffolds for articular cartilage tissue engineering. *Acta Biomater.*, 7, 12, 4187–94, 2011.
69. Ryu, J. *et al.*, Mussel-inspired polydopamine coating as a universal route to hydroxyapatite crystallization. *Adv. Funct. Mater.*, 20, 13, 2132–2139, 2010.
70. Xie, J. *et al.*, Controlled biomineralization of electrospun poly(ε-caprolactone) fibers to enhance their mechanical properties. *Acta Biomater.*, 9, 3, 5698–707, 2013.
71. Wu, C. *et al.*, Mussel-inspired porous SiO2 scaffolds with improved mineralization and cytocompatibility for drug delivery and bone tissue engineering. *J. Mater. Chem.*, 21, 45, 18300–18307, 2011.
72. Ding, Y. *et al.*, Modulation of protein adsorption, vascular cell selectivity and platelet adhesion by mussel-inspired surface functionalization. *J. Mater. Chem. B*, 2, 24, 3819–3829, 2014.
73. Rim, N.G. *et al.*, Mussel-inspired surface modification of poly(L-lactide) electrospun fibers for modulation of osteogenic differentiation of human mesenchymal stem cells. *Colloids Surf. B Biointerfaces*, 91, 189–97, 2012.
74. Xu, M. *et al.*, Mussel-inspired bioactive ceramics with improved bioactivity, cell proliferation, differentiation and bone-related gene expression of MC3T3 cells. *Biomater. Sci.*, 1, 9, 933–941, 2013.
75. Steeves, A.J. *et al.*, Evaluation of the direct effects of poly(dopamine) on the *in vitro* response of human osteoblastic cells. *J. Mater. Chem. B*, 4, 18, 3145–3156, 2016.
76. Lee, J.H. *et al.*, Guidance of *in vitro* migration of human mesenchymal stem cells and *in vivo* guided bone regeneration using aligned electrospun fibers. *Tissue Eng. Part A*, 20, 15–16, 2031–42, 2014.

77. Li, Y. *et al.*, Improving osteointegration and osteogenesis of three-dimensional porous Ti6Al4V scaffolds by polydopamine-assisted biomimetic hydroxyapatite coating. *ACS Appl. Mater. Interfaces*, 7, 10, 5715–5724, 2015.
78. Jia, L. *et al.*, Polydopamine-assisted surface modification for orthopaedic implants. *J. Orthop. Transl.*, 17, 82–95, 2019.
79. Yang, Z. *et al.*, Mussel-inspired coating of polydopamine directs endothelial and smooth muscle cell fate for re-endothelialization of vascular devices. *Adv. Healthc. Mater.*, 1, 5, 548–59, 2012.
80. Luo, R. *et al.*, *In vitro* investigation of enhanced hemocompatibility and endothelial cell proliferation associated with quinone-rich polydopamine coating. *ACS Appl. Mater. Interfaces*, 5, 5, 1704–14, 2013.
81. Ding, Y. *et al.*, Modulation of protein adsorption, vascular cell selectivity and platelet adhesion by mussel-inspired surface functionalization. *J. Mater. Chem. B*, 2, 24, 3819–3829, 2014.
82. Liu, M. *et al.*, Surface modification of zirconia with polydopamine to enhance fibroblast response and decrease bacterial activity *in vitro*: A potential technique for soft tissue engineering applications. *Colloids Surf. B Biointerfaces*, 136, 74–83, 2015.
83. Su, L. *et al.*, Strong antibacterial polydopamine coatings prepared by a shaking-assisted method. *Sci. Rep.*, 6, 24420, 2016.
84. Pashley, D.H. *et al.*, From dry bonding to water-wet bonding to ethanol-wet bonding. A review of the interactions between dentin matrix and solvated resins using a macromodel of the hybrid layer. *Am. J. Dent.*, 20, 1, 7–20, 2007.
85. Jee, S.E. *et al.*, Investigation of ethanol infiltration into demineralized dentin collagen fibrils using molecular dynamics simulations. *Acta Biomater.*, 36, 175–85, 2016.
86. Fang, H. *et al.*, Anti-proteolytic property and bonding durability of mussel adhesive protein-modified dentin adhesive interface. *Dent. Mater.*, 33, 10, 1075–1083, 2017.
87. Devarajan, S. *et al.*, Effect of polydopamine incorporated dentin adhesives on bond durability. *J. Adhes. Sci. Technol.*, 35, 1–14, 2020.
88. Zhou, Y.Z. *et al.*, Polydopamine-induced tooth remineralization. *ACS Appl. Mater. Interfaces*, 4, 12, 6901–10, 2012.
89. Hauser, D. *et al.*, From bioinspired glue to medicine: Polydopamine as a biomedical material. *Mater. (Basel)*, 13, 7, 120, 2020.
90. Ding, Y.H., Floren, M., Tan, W., Mussel-inspired polydopamine for biosurface functionalization. *Biosurf. Biotribol.*, 2, 4, 121–136, 2016.

5

Application of Electroactive Polymer Actuator: A Brief Review

Dillip Kumar Biswal

Krupajal Engineering College, Bhubaneswar, Odisha, India

Abstract

Studies have been directed for more sustainable materials that could really respond to specific stimulation while retaining its original form. When stimulated by mechanical, electrical, chemical, optical, or magnetic field, some polymers alter their structure or size. Electric energy seems to be the most appealing stimulating approach to electroactive polymers (EAP) and artificial muscles that can operate both as actuator and sensor in a timely manner. Electroactive polymers are outstanding alternatives despite their difficult manufacturing conditions because of their high strain rate, rapid responsiveness, reliability, and mechanical compatibility. Bioinspired characteristics are among the most appealing applications for EAP materials. Currently, there are a variety of EAP resources available. Some EAP materials can transfer mechanical strain to electrical signals, making them excellent energy harvesters and sensors. Efforts are being done around the world to develop EAP materials becoming thrusters, which includes developing their scientific and engineering foundation for the development, as well as comprehending their operation principles. These efforts have resulted in a significant understanding of the factors that influence their competence and persistence. In addition, efficient processing approaches for manufacturing, structuring, electroding, and analysis are being developed. This study explored different types of electroactive polymers, general enabling mechanisms that make advantage of their distinctive capabilities, the factors that determine the properties and different applications of electroactive polymers.

Keywords: Smart materials, electroactive polymers, bioinspired polymers, artificial muscles

Email: dillipkumarbiswal@gmail.com

Inamuddin, Tariq Altalhi and Ashjan Alrogi (eds.) Biomimicry Materials and Applications, (127–146)

5.1 Introduction

Electroactive polymers (EAPs) are polymeric composite whose geometry and specifications change when exposed to an electric potential [1]. They show a lot of strain yet producing a greater force with the application low voltage input [2]. EAPs are being used as actuators [1] and sensors [3, 4] because of their special trait and potential. When triggered by an applied electric field, EAP actuators produce a mechanical response, whereas EAP sensors generate an electrical current when stimulated mechanically. Ionic electroactive polymers (iEAPs) is an ionic-conductive membrane placed between two conducting electrodes that bend due to the movement of electrolyte ions into electrodes within the membrane [5]. The field's significant advancements are piquing the interest of researcher's and engineers from a variety of fields [6, 7].

Several EAP actuators are very much in the initial phases of development, and it will require more development before they can be used in mass-produced applications. Appropriate manufacturing, processes are indeed being designed to optimize their actuation performance and robustness. In addition, methodologies for identifying and recording their responses devised [8]. Developers are continuously on the lookout for unique specific applications for these materials so how they can be mass-produced on a regular basis.

5.2 Chronological Summary of the Evolution of EAP Actuator

Roentgen was the first to undertake a recorded experiment with EAP materials in the year 1880 [9]. In 2013, Keplinger *et al.* showed how this electrode-free approach could be utilized to elicit significant strains in an actuator by repeatedly sprinkling of electrical charges onto a dielectric [10]. The identification of electret is a great milestone in the development of electroactive polymer [11]. Electrets are electroactive in the sense that they deform with the application of electric potential and generate an electric potential as when distorted. Electrets' applicability has been confined to sensors due to the little strain generated by the electric field.

Katchalsky *et al.* [12] are the ones who are credited with being the ones who initially thought of the concept of responsive gels. Hokkaido University in Japan is credited with carrying out research that was both significant and ground-breaking [13]. This research focused on responsive gels and the electrochemical activation of such gels. Recent research has shown that some new kinds of polymer gels can stretch a lot with just a small amount of activation voltage [14].

Following the discovery of electrets, scientists studied piezoelectric biopolymers [15] and discovered that PVDF had strong piezoelectric activity [16]. Following the invention of PVDF as an electroactive polymer, a comprehensive search for additional polymer systems with considerable responsiveness was undertaken. PVDF was the subject of intensive research in the 1970s and 1980s, mostly in an effort to increase performance and find applications [17]. PVDF is mostly used as sensors and ultrasonic wave transducers due to its low strain [18]. Efficacy in producing novel EAP materials was largely documented in the 1990s [19–23].

Bar-Cohen began his study in the field of EAP in 1995 and immediately realized that forming global cooperation was important in order to speed up the evolution of EAP materials and contribute significantly to functional actuators [24]. As a result, he established a number of information-sharing venues, such as "SPIE annual EAP actuators and devices (EAPAD) Conference." In order to encourage global growth toward the capabilities of EAP substances, the principal author issued an "Arm Wrestling Challenge" in 1999 "http://ndeaa.jpl.nasa.gov/nasa-nde/lommas/eap/EAP-armwrestling.htm."

5.3 Electroactive Polymer Actuators Groups

Electroactive polymers are mainly categorized into two classes. Ionic EAP and electronic EAP are two types of electroactive polymer depending on how they are synthesized. As outlined in Figure 5.1, both of these categories could be further subdivided into multiple branches.

Dielectric elastomers, for example, are electronic EAPs that inflate mechanically in response to electrical charge variations. Similarly, a potential difference can activate ionic EAPs by inducing deformation due to ion mobility. Ionic EAPs, on the other hand, imply lower energy to stimulate compared to electronic EAPs, and hence have attracted considerable attention in the research [25]. Electronic EAPs have advanced tremendously, and a variety of electric EAPs are now available on the market. According to statistics, the electroactive polymer market was valued at US$ 4,475.00 million in 2020 and is predicted to increase at a CAGR of 7.3% from 2020 to 2028, to reach US$ 1,340.3 million [26]. Table 5.1 summarizes the strengths and weaknesses of the different EAP material categories.

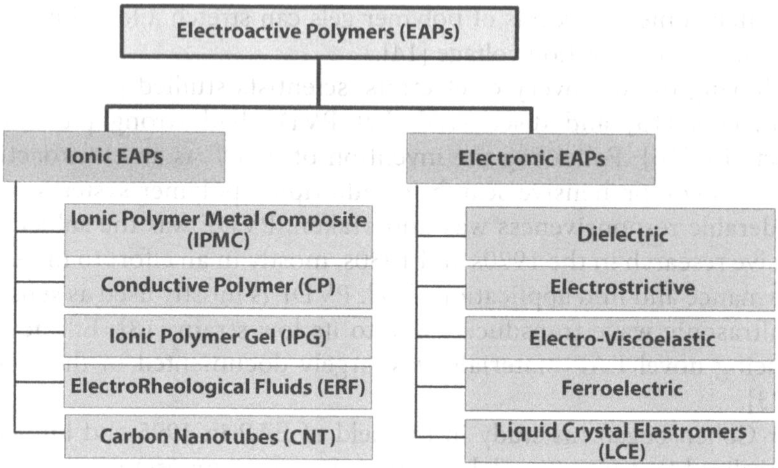

Figure 5.1 Types of electroactive polymers.

5.3.1 Ionic Electroactive Polymers

Ionic electroactive polymers (iEAPs) are by far the most prevalent form of electroactive polymers. They require less voltage to activate than other EAPs since they are driven by ion diffusion. These ionic EAPs have a steep bending angle, which allows them to more accurately simulate muscle activity. Figure 5.2 depicts the actuator structure of an ionic EAP, which includes an ion exchange membrane sandwiched between two conducting electrodes. When an electric potential is applied to the ionic EAP, hydrated cations migrate toward the negative terminal, generating a localized volume change causing bending deformation as shown in Figure 5.2. In the ionic EAP, this volume shift causes a distinctive bending motion [27].

The findings of Hong *et al.* revealed that an actuator made of drop-cast PEDOT: PSS on Nafion may bend sharply above 90^0 [28]. Furthermore, whenever the ionic EAP is deformed mechanically, an electric signal is generated, allowing it to operate as both an actuator and a sensor [29].

Ionic EAPs can be made in a variety of ways using various materials. Ionic EAPs can be classified into a variety of groups based on these pathways, including ionic polymer-metal composites, conductive polymers, carbon nonotubes, ionic polymer gels, and electrorheological fluids [30]. It is crucial to remember that these categories are not fixed. Instead, as their characteristics are researched and understood, they are frequently joined into novel ionic EAPs.

Table 5.1 Summary of the benefits and drawbacks of the two primary EAP groups.

EAP	Advantages	Disadvantages
Ionic	• Large bending deformation is being obtained. • Low electric potential is required for activation. • Featuring bi-directional actuation that is dependent on the polarity of the applied electric potential.	• With the exception of Carbon Nanotubes and Conductive Polymers, ionic EAPs do not hold strain in a DC electric potential. • Response time is slow. • Bending EAPs results a small actuation force. • It is difficult to fabricate these materials with repeatable content, with the exception of Conductive Polymers. • In fluid frameworks, they endure electrolysis at > 1.23 V. • They involve an aqueous system that needs electrolyte content to be maintained when working in open atmosphere. • Low electromechanical coupling productivity.
Electronic	• It can function for a longer period of time under room conditions. • A rapid response (in milliseconds). • Under DC activation, it can hold strain. • Is capable of generating relatively large actuation forces.	• Requires a strong field, which might result in high voltages. • The electrostriction requires monopolar actuation regardless of voltage polarity.

5.3.2 Electronic Electroactive Polymers

In the last few decades, electronic EAP has garnered a lot of interest in the scientific community. Unlike ionic EAP, electronic EAP do not need an electrolyte environment or ion mobility to function [31]. As a result, the

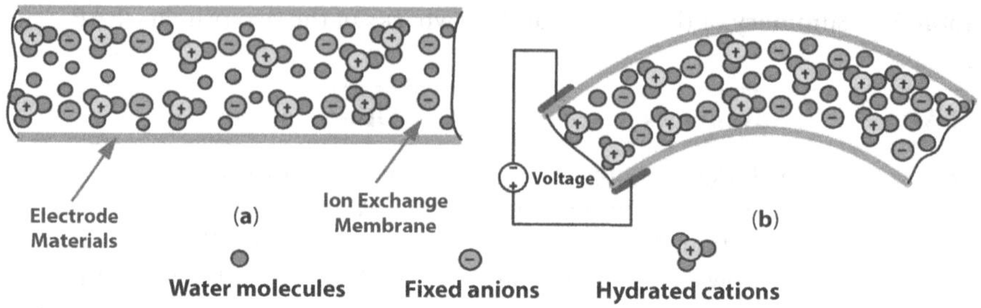

Figure 5.2 Structure of ionic EAP.

electronic EAP are getting more and more popular, thanks to a growing interest in lightweight, biocompatible, and easy-to-manufacture materials [32]. Various types of electronic EAPs have been created so far for a variety of purposes. Electric EAPs such as dielectric, electrostrictive paper, liquid crystal, ferroelectric, electrostrictive graft, and electroviscoelastic, are well known in the literature [1, 32, 33].

5.4 Application of Electroactive Polymer Actuators

Electroactive polymers, in particular, have been intensively explored as actuators due to their unique ability to transform electric potential into mechanical force. In cases when substantial mechanical elements are not desirable, EAPs can be quite beneficial. Actuators made of electroactive polymers are also a single component, rather than a collection of pieces that might rub together and create wear. Electroactive polymers are silent, and their use might greatly reduce friction and wear compared with traditional actuators [34, 35]. Additionally, adaptability permits EAPs to transform their traits in response to changing circumstances [36]. Because of these characteristics, EAP actuators can be used in a variety of applications where standard actuators are ineffective. Electroactive polymer-based coating has also been a feasible solution for increasing performance of the product in recent years. By treating a base material with EAPs, a number of benefits become available. These benefits include increased electrochemical capacitance, reduced electric resistance, reduced friction and wear, corrosion resistance, oxide surface stability and biomedical applications [37]. As a consequence, EAPs have been successfully used to fabricate a wide range of devices in fields like engineering and medicine. The next section discusses several intriguing EAP applications.

5.4.1 Soft Robotic Actuator Applications

Ionic electro-active polymers (EAPs) have recently received a lot of interest because they have a low actuation electric potential, a high strain, a low density, a noiseless operation, and flexibility comparable to natural muscle [38, 39]. Their adaptability and flexibility enable innovative robotic movements, including a serpentine-moving snake-shaped robot [40]. With their adaptability enables the creation of sophisticated architectures, like mimicked human fingertip. Because most EAPs can indeed be adapted into actuation devices, practically all EAPs can be employed in robotics based on unique requirements. However, a few varieties stand out as the most commonly used. Relatively large bending deflection, light in weight, ease of manufacture, and versatility, dielectric elastomers may be extremely beneficial in certain robotic systems [41]. Since conductive polymer actuators are often biocompatible and could be manufactured in micro/nanodimensions, they have potential to be beneficial in biomedical field [42]. Depending on the specific situations in which the actuator is meant to be employed, various alternative types of EAPs are appropriate. In the domain of acoustical technology, some EAP actuators have even been employed to attenuate sound. Noise-cancelling headphones, transducers, and resonators have indeed used EAP actuators [43–45]. Electrical EAP actuators, on the other hand, have high activation electric potential that could be hazardous to users. Another difficulty is dealing with dielectric breakdown, which really is undesirable in artificial muscle [46, 47]. Certain ionic EAPs can be utilized to produce synthetic muscles owing to their bending deflection that could be used to accomplish translational movement well with right procedures [48]. Tribological considerations are important irrespective of whatever type of electroactive polymers are utilized as an artificial muscle because of the physical discomfort which friction may inflict in the patients psyche [49]. Additionally, EAP coating has also been proven to be effective in the development of adaptable artificial muscles. Ebadi *et al.* coated polyurethane nanofibers with EAP and produced an angular displacement of 48° to 225° [50]. This demonstrates a clear edge, particularly when flexible structures are seen as muscles.

5.4.2 Underwater Applications

Electroactive polymer actuators can also be employed in underwater applications for clatter-free movement. Conventional methods of propulsion have been found to have a number of harmful effects on marine species [51]. EAP actuators could be used as a low-noise substitute. Electroactive polymer actuators can also be miniature and replicate aquatic life movement, which would be important in oceanic engineering. Because many variants of ionic EAPs respond well in aqueous environments, they are especially effective underwater applications. A "turtle-like robot" with inductive charging for aquatic use has been designed, showcasing the breadth of IPMC applications [52]. IPMCs are extremely deformable at even low electric potential, soft & flexible and are especially favorable underwater because they do not suffer from dehydration effect. It has already been established that multilayered IPMC actuator is more efficient assuring underwater response compared to a single-layered IPMC actuator [53]. In an underwater environment, conducting polymer and ionic gel actuators offer similar benefits, while IPMCs have received greater attention [42, 54].

5.4.3 Aerospace Applications

Electroactive polymer materials' electromechanical characteristics allow them to work as both actuators and sensors in the aerospace sector. Future NASA missions will require properties, such as low density, compactness, and low power consumption materials, i.e., EAP materials [55]. In extensive research carried out by Punning *et al.* [56], the researchers tested and contrasted the performance of a number of ionic EAP actuators that were developed specifically for use in aerospace applications. The findings of the study indicate that EAP actuators are completely unaffected by the X-ray radiation and ionizing gamma that is found in space. The thickness fraction of the composite sheets serves as an important influence in actuator deformation and holds tremendous promise for future morphing applications [57]. To diminish the separation bubble, De Mauro *et al.* used an EAP actuator to regulate the flow over an airfoil [58]. EAP materials have assumed the position of electric motors in the structure due to qualities, such as flexibility, compactness, quick reaction speed, and low energy consumption [59]. Li *et al.* introduced an artificial muscle that can opens up the door towards efficient design and low-cost manufacture of actuation to enormous deployable structures for space research [60]. IPMC is employed as a dust wiper on the infrared camera lens of a nano-rover in the MUSES-CN Mission [61–63]. The European Space Agency recently sought to launch space-based initiatives to research the production procedures and controlling capacity of seamless integration smart materials at the initial stages of space systems [56, 64]. Stanford Research Institute has also investigated dielectric actuators for shape control of lightweight space mirrors [65]. There have been cases when EAPs have been utilized in micro air vehicles (MAVs) that are equipped with wings that are able to flap [66–73]. EAP is the best choice for usage as an actuator in micro-robots to achieve flapping flight owing to their small density and high displacement rate [74–76].

The researchers created a prototype of a paper glider that had printed transistors and actuators that were lightweight, mechanically flexible, and mechanically adaptable [77]. This combination was developed with the intention of demonstrating electrically controlled glide path changing in an unmanned aerial vehicle (UAV) system that was lightweight and disposable. EAPs are used as "wind turbine," "rotor blades," "edge flight ailerons," and "bumps" in transonic flow [78]. Transonic flow is the part of an airplane's operation that happens when it is flying at a steady speed.

5.4.4 Energy Harvesting Applications

Ionic electroactive polymers (iEAPs) may create electrical energy during bending deformations, implying that they have a high potential for building energy harvesters that draw power from vibrant pulsating situations. Electroactive polymers are currently being utilized as the next-generation intelligent material in the majority of energy harvesting systems. This is due to the fact that electroactive polymers have roughly 10 times the energy density of conventional energy conversion elements. Kweon *et al.* tested the energy-harvesting capabilities by altering the porosity of the electrolyte membrane [79]. The prospect of building a generator is being investigated as a result of the emergence of energy harvesting technologies [80]. Increased electrochemical capacitance is yet another applicability of EAP-based coating. There is a possibility that this coating will alter the electrodes in electrochemical energy storage systems [81].

5.4.5 Healthcare and Biomedical Applications

Because of their flexibility, substantial deformation, and biocompatibility, IPMCs have prospective applications in medical engineering [82]. By directing the movement of the laser beam in a number of different directions, an IPMC actuator with four electrodes can improve the accuracy of an optical fibre actuator that is used in micro-ophthalmic surgery [83]. Feng and Chen established a micro machined method for producing IPMC devices for laparoscopic use [84]. In biomedical active catheter systems, cylindrical IPMC actuators allow for motion with multiple degrees of freedom and provide fine control over the placement of the tip [82, 85–87]. The cylindrical IPMC was employed by the Eamex Corporation for the manufacture of brain catheters [88]. In their creation of an artificial heart, Shahinpoor *et al.* [89] included an ionic polymer metal composite actuator as a supplementary diastolic device. An IPMC was used to regulate the motion of tiny robots in capsules throughout the body [90].

A recent development in EAP-based coatings has been recognized as a significant step forward in the field of medication administration into the bloodstream. Patients can be provided with the needed doses of drugs based on polymers in a controlled fashion. It was possible for the EAP to give the anti-inflammatory medicine dexamethasone phosphate to the location in question. These biodegradable electroactive polymers help to keep people safe while still being environmentally friendly. This technique has a number of advantages, the most important of which is that it enables the medication to be administered at more opportune periods, hence minimizing the risk of adverse effects [91]. This method makes very limited use of electricity and operates at a low voltage in order to prevent harm to human beings. Hardy *et al.* studied the use of oligoaniline-based electroactive blocks connected to polycaprolactone blocks in drug delivery applications [92]. The progress that is being made right now has a lot of potential, and more research is being done to make sure that applications of this kind are reliable [93].

Human cells can be reproduced and stuck to the use of electroactive polymers as biomaterials [94–97]. Hamiltonian model is used to create the best electroactive polymer for a medical endoscope application [98]. A wide range of pharmacologically active chemicals are being studied in order to determine drug release characteristics against a variety of polymers that are either ionic in nature or have the property of intrinsic conductivity [99-100]. The best technique of drug loading is determined by the properties of the drug itself, whereas the best way of drug release is determined mostly by the ionic character of the molecule [101–104]. Active and regulated medication administration on demand can be facilitated by electroactive materials that respond within biological circumstances [105].

Using a conducting polymer known as poly (3, 4-ethylenedioxypyrrole) (PEDOP) that is loaded with an antibiotic drug known as tetracycline, an antibacterial and neuroprotective coating was developed. This coating was designed to prevent the spread of microorganisms (Tc). It has been demonstrated that PEDOP/Tc can perform as an effective brain electrode coating capable of enhancing charge transfer while also exhibiting bi-functional biological properties that differ between eukaryotic and prokaryotic cells [106]. Puiggali *et al.* described the creation of conductive polymer NPs as therapeutic Cys-Arg-Glu-Lys-Ala (CREKA) carriers for regulated dosage administration via electric stimulation [107].

5.4.6 Shape Memory Polymer Applications

After being heated past a certain transition temperature, a substance is said to exhibit the phenomenon known as the shape memory effect, in which the substance returns to its original form and size. Recent studies [108, 109] have demonstrated that electroactive shape memory polymers, also known as EASMPs, are ideally suited for use with remotely programmable actuators in situations when it is not possible to apply heat directly to the material. EAPs that have a high glass transition temperature can maintain their shape memory capabilities even at higher temperatures. It was determined that the EASMPI had a recovery stress of 40.1 MPa, which meant that it was capable of putting forth mechanical effort [108]. Electroactive shape memory polyimide (EASMPI) exhibits both electroactive and shape memory properties. Ground carbon fibres (GCFs) were incorporated into the polyimide matrix to create EASMPI. Shape memory capabilities can be provided at greater temperatures thanks to EAPs with a high glass transition temperature [108, 110, 111].

5.4.7 Smart Window Applications

A number of knowledgeable individuals have also been conducting research on the cutting-edge technology that is the smart window. The researchers, Xie *et al.* [112] used a material known as bistable electroactive polymer (BSEP) to create smart windows. These windows can change from an opaque to a transparent state, as well as between semi-crystalline and amorphous states, based on whether the material is warmed or allowed to cool at a specific temperature. It has been shown that the resulting smart window can change the amount of solar radiation that gets through by 70.2% while also changing the amount of emission radiation that gets through by 80.4%.

5.4.8 Wearable Electronics Applications

In the realm of wearable electronics, EAP actuators are also widely used. They could really conform to the contours of human body segments since these actuators are flexible in nature. Furthermore, they may be manufactured as very small devices, allowing them to be used in more sophisticated applications. A touch display for a living person fingertip was developed by Koo *et al.*, which consists of a series of dielectric elastomer-based haptic simulators [113]. It is even conceivable to design an EAP-based fabric with the ability to vary density depending on the activity [114]. Wearable electroactive polymer may be used to gather mechanical energy since when they are distorted, they generate an electric charge. Because electronic EAPs are simpler to use in the open environment than ionic EAPs, they seem to be the most often employed form of EAP for these purposes. Ionic electroactive polymer, on the other hand, make preferable for human usage owing to its low activation electric potential. Due to its outstanding characteristics, PVDF is almost like an electrical EAP which has gain recognition for wearables [115]. IPMC sensors were used to create a data glove that captured the bending action of a human finger [116]. Currently, the Robert Bosch Corporation in Singapore is exploring the potential of electrodynamic polymeric actuator and related applicability in flexible wearable technology.

5.5 Conclusion

Electroactive polymers are a form of functional polymer that has attracted a lot of buzz in the fields of engineering, medicine, and other fields. EAPs can be manufactured with specific properties in a variety of ways, depending on the application. Some essential qualities and properties of electroactive polymers have been elucidated in this chapter, and some contemporary applications in robotics, aerospace, energy harvesting, and biomedical have been discussed. The implementation of this technology stands to boost safety and benefit users considering the noiseless activation, substantially reduced voltage demands compared to standard devices. EAPs have the potential to be used as soft robotic actuators because they do not require integration into rigid components to work. Noiseless propulsion has the potential to reduce the negative effects of motorized underwater machines on aquatic life in underwater applications. It is clear that EAPs have a lot of promise for use in a variety of sectors. Exploration of EAPs and their features and uses is currently ongoing, and there are numerous prospects and research areas for EAPs.

References

1. Bar-Cohen, Y., *Electroactive polymer (EAP) actuators as artificial muscles: Reality, potential, and challenges*, pp. 1–765, SPIE Press, Bellingham, Washington, 2004.
2. Popović, M.B., *Biomechanics and robotics*, Taylor & Francis Group, Boca Raton, 2013.
3. Wang, T., Farajollahi, M., Choi, Y.S., Lin, I.T., Marshall, J.E., Thompson, N.M., Smoukov, S.K., Electroactive polymers for sensing. *Interface Focus*, 6, 20160026, 2016.
4. Miki, H., Matsui, G., Kanda, M., Tsuchitani, S., Fabrication of microstructure array directly on β-phase poly (vinylidene fluoride) thin film by O_2 reactive ion etching. *J. Micromech. Microeng.*, 25, 035026, 2015.

5. Punning, A., Kim, K.J., Palmre, V., Vidal, F., Plesse, C., Festin, N., Spinks, G., Ionic electroactive polymer artificial muscles in space applications. *Sci. Rep.*, 4, 6913, 2014.
6. Kim, K.J., Last twenty-five years of effort in developing fabrication-methods of IPMCs, paper no. 10594-9. *Proceedings of the EAPAD Conference, SPIE Smart Structures and Materials Symp.*, Held in Denver, CO, 2018.
7. Madden, J.D.W., B25 years of conducting polymer actuators: History, mechanisms, applications, and prospects, paper no. 10594-2. *Proceedings of the EAPAD Conference, SPIE Smart Structures and Materials Symp.*, Held in Denver, CO, 2018.
8. Carpi, F., Anderson, I., Bauer, S., Frediani, G., Gallone, G., Gei, M., Graaf, C., Jean-Mistral, C., Kaal, W., Kofod, G., Standards for dielectric elastomer transducers. *Smart Mater. Struct.*, 24, 105025, 2015.
9. Roentgen, W.C., About the changes in shape and volume of dielectrics caused by electricity, section III, in: *Annual Physics and Chemistry Series*, vol. 11, G. Wiedemann, (Ed.), pp. 771–786, John Ambrosius Barth Publisher, Leipzig, 1880.
10. Keplinger, C., Sun, J.Y., Foo, C.C., Rothemund, P., Whitesides, G.M., Suo, Z., Stretchable, transparent, ionic conductors. *Science*, 341, 984, 2013.
11. Eguchi, M., On the permanent electret. *Philos. Mag.*, 49, 178, 1925.
12. Katchalsky, A. and Zwick, M., Mechano chemistry and ion exchange. *J. Polym. Sci.*, 16, 221, 1955.
13. Osada, Y. and Kishi, R., Reversible volume change of micro particles in an electric field. *J. Chem. Soc.*, 85, 655, 1989.
14. Osada, Y., Chemical valves and gel actuators. *Adv. Mater.*, 3, 107, 1991.
15. Fukada, E., Piezoelectricity of wood. *J. Phys. Soc. Jpn.*, 10, 149, 1955.
16. Kawai, H., Piezoelectricity of poly (vinylidene fluoride). *Jpn. J. Appl. Phys.*, 8, 975, 1969.
17. Furukawa, T., Piezoelectricity and pyroelectricity in polymers. *IEEE Trans. Electr. Insul.*, 24, 375, 1989.
18. Bauer, S. and Bauer, F., Piezoelectric polymers and their applications, in: *Piezoelectricity Evolution and Future of a Technology Series: Springer Series in Materials Science*, vol. 114, Heywang, W., Lubitz, K., Wersing, W. (Eds.), pp. 157–177, 2008.
19. Baughman, R.H., Conducting polymer artificial muscles. *Synth. Met.*, 78, 339–353, 1996.
20. Baughman, R.H., Cui, C., Zakhidov, A.A., Iqbal, Z., Basrisci, J.N., Spinks, G.M., Wallace, G.G., Mazzoldi, A., De Rossi, D., Rinzler, A.G., Jaschinski, O., Roth, S., Kertesz, M., Carbon nanotube actuators. *Science*, 284, 1340–1344, 1999.
21. Lovinger, A., Ferroelectric polymers. *Science*, 220, 1115, 1983.
22. Nalwa, H.S. (Ed.), *Ferroelectric polymers – chemistry, physics*, and applications, Marcel Dekker, Inc., New York, 1995.
23. Oguro, K., Kawami, Y., Takenaka, H., Bending of an ion-conducting polymer film-electrode composite by an electric stimulus at low voltage. *Trans. J. Micromach. Soc.*, 5, 27, 1992.

24. Bar-Cohen, Y. (Ed.), Proceedings of the first SPIE's electroactive polymer actuators and devices (EAPAD). *Conf., Smart Structures and Materials Symposium*, vol. 3669, pp. 1–414, 1999.
25. Neuhaus, R., Zahiri, N., Petrs, J., Tahouni, Y., Siegert, J., Kolaric, I., Dahy, H., Bauernhansl, T., Integrating ionic electroactive polymer actuators and sensors into adaptive building skins: Potentials and limitations. *Front. Built Environ.*, 6, 95, 2020.
26. Electroactive polymer market size, share & analysis, by product type (conductive polymer, inherently conductive polymer, inherently dissipative polymer, and others), by activation mode, by application, and by region, forecast to 2028, 2022, https://www.researchandmarkets.com/reports/539304/electroactive-polymer-market-size-share-and#rela3-5025764.
27. Jo, C., Pugal, D., Oh, I.-K., Kim, K.J., Asaka, K., Recent advances in ionic polymer–metal composite actuators and their modelling and applications. *Prog. Polym. Sci.*, 38, 1037, 2013.
28. Hong, W., Almomani, A., Chen, Y., Jamshidi, R., Montazami, R., Soft ionic electroactive polymer actuators with tunable non-linear angular deformation. *Materials*, 10, 664, 2017.
29. Palmre, V., Brandell, D., Mäeorg, U., Torop, J., Volobujeva, O., Punning, A., Johanson, U., Kruusmaa, M., Aabloo, A., Nanoporous carbon-based electrodes for high strain ionomeric bending actuators. *Smart Mater. Struct.*, 18, 095028, 2009.
30. Kim, O., Kim, S.J., Park, M.J., Low-voltage-driven soft actuators. *Chem. Commun.*, 54, 4895–4904, 2018.
31. Wang, T., Farajollahi, M., Choi, Y.S., Lin, I.T., Marshall, J.E., Thompson, N.M., Kar-Narayan, S., Madden, J.D.W., Smoukov, S.K., Electroactive polymers for sensing. *Interface Focus*, 6, 20160026, 2016.
32. Araromi, O.A., Conn, A.T., Ling, C.S., Burgess, S.C., Vaidyanathan, R., A novel fabrication set-up for the flexible production of silicone based EAP "artificial muscle" actuators. *Des. Nat. V*, 92, 289, 2010.
33. Pelrine, R.E., Kornbluh, R.D., Joseph, J.P., Electrostriction of polymer dielectrics with compliant electrodes as a means of actuation. *Sens. Actuators A Phys.*, 64, 77, 1998.
34. Punning, A., Kim, K.J., Palmre, V., Vidal, F., Plesse, C., Festin, N., Maziz, A., Asaka, K., Sugino, T., Alici, G., Ionic Electroactive polymer artificial muscles in space applications. *Sci. Rep.*, 4, 6913, 2014.
35. Menezes, P.L., Nosonovsky, M., Ingole, S.P., Kailas, S.V., Lovell, M.R. (Eds.), *Tribology for scientists and engineers*, Springer, New York, NY, USA, 2013.
36. Mun, S., Yun, S., Nam, S., Park, S.K., Park, S., Park, B.J., Lim, J.M., Kyung, K.U., Electro-active polymer based soft tactile interface for wearable devices. *IEEE Trans. Haptics*, 11, 15, 2018.
37. Lyons, M.E.G., *Electroactive polymer electrochemistry: Part 1: Fundamentals*, Springer Science & Business Media, Berlin, Germany, 2013.

38. Pugal, J.C., Oh, D.I.K., Kim, K.J., Asaka, K., Recent advances in ionic polymer-metal composite actuators and their modeling and applications. *Prog. Polym. Sci.*, 38, 1037, 2013.
39. Mukai, K., Asaka, K., Sugio, T., Kiyohara, K., Takeuchi, I., Terasawa, N., Futaba, D.N., Hata, K., Fukushima, T., Aida, T., Highly conductive sheets from millimeter-long singlewalled carbon nanotubes and ionic liquids: Application to fastmoving, low-voltage electromechanical actuators operable in air. *Adv. Mater.*, 21, 1582, 2009.
40. Ahmed, M. and Billah, M.M., Smart material-actuated flexible tendon-based snake robot. *Int. J. Adv. Robot. Syst.*, 13, 89, 2016.
41. Jung, K., Koo, J.C., Lee, Y.K., Choi, H.R., Artificial annelid robot driven by soft actuators. *Bioinspir. Biomim.*, 2, S42, 2007.
42. Hines, L., Petersen, K., Lum, G.Z., Sitti, M., Soft actuators for small-scale robotics. *Adv. Mater.*, 29, 1603483, 2017.
43. R.N. Hitchcock, D.W. Humphreys, A.H. Muir, A. Zarrabi, Audio devices with electroactive polymer actuators noise cancellation. Google Patents W0209500573A2, 26 June 2014.
44. Heydt, R.P., Pelrine, R.E., Kornbluh, R.D., Bonwit, N.A., Eckerle, J.S., Compliant electroactive polymer transducers for sonic applications. *Acoust. Soc. Am. J.*, 130, 1078, 2011.
45. Abbad, A., Rabenorosoa, K., Ouisse, M., Atalla, N., Adaptive Helmholtz resonator based on electroactive polymers: Modeling, characterization, and control. *Smart Mater. Struct.*, 27, 105029, 2018.
46. Qiu, Y., Zhang, E., Plamthottam, R., Pei, Q., Dielectric elastomer artificial muscle: Materials innovations and device explorations. *Acc. Chem. Res.*, 52, 316, 2019.
47. Palumbo, F., Wen, C., Lombardo, S., Pazos, S., Aguirre, F., Eizenberg, M., Hui, F., Lanza, M., A review on dielectric breakdown in thin dielectrics: Silicon dioxide, high-k, and layered dielectrics. *Adv. Funct. Mater.*, 30, 1900657, 2020.
48. Fannir, A., Temmer, R., Nguyen, G.T.M., Cadiergues, L., Laurent, E., Madden, J.D.W., Vidal, F., Plesse, C., Linear artificial muscle based on ionic electroactive polymer: A rational design for open-air and vacuum actuation. *Adv. Mater. Technol.*, 4, 1800519, 2019.
49. Klontzas, M.E., Akoumianakis, I.D., Vagios, I., Karantanas, A.H., MR imaging findings of medial tibial crest friction. *Eur. J. Radiol.*, 82, e703, 2013.
50. Ebadi, S.V., Fashandi, H., Semnani, D., Rezaei, B., Fakhrali, A., Electroactive actuator based on polyurethane nanofibers coated with polypyrrole through electrochemical polymerization: A competent method for developing artificial muscles. *Smart Mater. Struct.*, 29, 045008, 2020.
51. McCormick, M.I., Fakan, E.P., Nedelec, S.L., Allan, B.J.M., Effects of boat noise on fish fast-start escape response depends on engine type. *Sci. Rep.*, 9, 6554, 2019.

52. Sun, Q., Han, J., Li, H., Liu, S., Shen, S., Zhang, Y., Sheng, J., A miniature robotic turtle with target tracking and wireless charging systems based on IPMCs. *IEEE Access*, 8, 187156, 2020.
53. Swarrup, J.S., Ganguli, R., Madras, G., Studies to improve the actuation capability of low-frequency IPMC actuators for underwater robotic applications. *ISSS J. Micro Smart Syst.*, 8, 41, 2019.
54. Li, M., Guo, S., Guo, J., Hirata, H., Ishihara, H., Development of a biomimetic underwater microrobot for a father–son robot system. *Microsyst. Technol.*, 23, 849, 2017.
55. Bar-Cohen, Y., Electroactive polymers as artificial muscles-capabilities, potentials and challenges, in: *Handbook on Biomimetics*, vol. 8, Y. Osada (Ed.), pp. 188–196, NTS Inc., Tokyo, Japan, 2000.
56. Punning, A., Kim, K.J., Palmre, V., Ionic electroactive polymer artificial muscles in space applications. *Sci. Rep.*, 4, 6913, 2014.
57. Ren, K., Bortolin, R.S., Zhang, Q., An investigation of a thermally steerable electroactive polymer/shape memory polymer hybrid actuator. *Appl. Phys. Lett.*, 108, 062901, 2016.
58. De Mauro, E.P., Dell'Orso, H., Zaremski, S., Control of laminar separation bubble on NACA 0009 airfoil using electroactive polymers. *AIAA J.*, 53, 2270, 2015.
59. Madden, J.D., Mobile robots: Motor challenges and materials solutions. *Science*, 318, 1094, 2007.
60. Li, S., Vogt, D.M., Rus, D., Fluid-driven origami inspired artificial muscles. *Proc. Natl. Acad. Sci.*, 114, 13132, 2017.
61. Ghamsari, A.K., *Bucky gel actuator for morphing applications*, KN Toosi University of Technology, 2012.
62. Starek., J.A., Acxıkmesxe, B., Nesnas, I.A., Spacecraft autonomy challenges for next-generation space missions, in: *Advances in Control System Technology for Aerospace Applications*, E. Feron (Ed.), pp. 1–48, Springer, Berlin, 2016.
63. Tsumaki, Y., Akaike, T., Kazama, R., Environment driven rover for asteroid exploration, in: *Proceedings of the 11th International Symposium on Artificial Intelligence, Robotics and Automation in Space*, Turin, Italy, 4–7 September 2012, IEEE, New York, 2012.
64. Feranec, J., Soukup, T., Hazeu, G., *European landscape dynamics: CORINE land cover data*, CRC Press, Boca Raton, FL, 2016.
65. Carpi, F., Sommer-Larsen, P., De Rossi, D., Electroactive polymers: New materials for spacecraft structures, in: *Spacecraft Structures, Materials and Mechanical Testing*, Noordwijk, 10–12 May 2005.
66. Anderson, M.L., Design and control of flapping wing micro air vehicles (DTIC document), 2011.
67. Anderson, M.L., Sladek, N.J., Cobb, R.G., Design, fabrication, and testing of an insect-sized MAV wing flapping mechanism, in: *49th AIAA Aerospace Sciences Meeting*, Orlando, FL, 4–7 January, 2011, AIAA, Reston, VA, p. 2549, 2011.

68. Baker, N.S., Hayes, M., Smith, D., Design of a flapping wing micro air vehicle actuation system, in: *Proceedings of the 2012 ASEE North-Central Section Conference*, Ada, OH, 23–24 March, 2012.
69. Burgess, S., Ling, C., Conn, A., Development of a novel electro active polymer (EAP) actuator for driving the wings of flapping micro air vehicle. *Computer Aided Optimum Design in Engineering*, vol. 11, p. 207, 2009.
70. Conn, A., Burgess, S., Hyde, R., From natural flyers to the mechanical realization of a flapping wing micro air vehicle, in: *2006 IEEE International Conference on Robotics and Biomimetics*, Kunming, China, 17–20 December, IEEE, New York, pp. 439–444, 2006.
71. Duggan, S.J., *An experimental investigation of flapping wing propulsion for micro air vehicles*, Naval Postgraduate School, Monterey, CA, 2000.
72. Pawlowski, K., Belvin, H., Raney, D., Electrospinning of a micro-air vehicle wing skin. *Polymer*, 44, 1309, 2003.
73. Sun, J., Guan, Q., Liu, Y., Morphing aircraft based on smart materials and structures: A state-of-the-art review. *J. Intell. Mater. Syst. Struct.*, 27, 2289, 2016.
74. Kim, K.J. and Tadokoro, S., *Electroactive polymers for robotic applications: Artificial muscles and sensors*, p. 291, London, Springer, 2007.
75. Schenato, L., Campolo, D., Sastry, S., Controllability issues in flapping flight for biomimetic micro aerial vehicles (MAVs), in: *Proceedings of the 42nd IEEE Conference on Decision and Control*, Maui, HI, 9–12 December, IEEE, New York, pp. 6441–6447, 2003.
76. Yan, J., Avadhanula, S., Birch, J., Wing transmission for a micromechanical flying insect. *J. Micromechatron.*, 1, 221, 2001.
77. Grau, G., Frazier, E.J., Subramanian, V., Printed unmanned aerial vehicles using paper-based electroactive polymer actuators and organic ion gel transistors. *Microsyst. Nanoeng.*, 2, 16032, 2016.
78. Villa, T.F., Gonzalez, F., Miljievic, B., An overview of small unmanned aerial vehicles for air quality measurements: Present applications and future prospectives. *Sensors*, 16, E1072, 2016.
79. Kweon, B.C., Sohn, J.S., Ryu, Y., Cha, S.W., Energy harvesting of ionic polymer-metal composites based on microcellular foamed nafion in aqueous environment. *Actuators*, 9, 71, 2020.
80. Chen, Z., Hedgepeth, D., Tan, X., A nonlinear, control-oriented model for ionic polymer–metal composite actuators. *Smart Mater. Struct.*, 18, 055008, 2009.
81. Abalyaeva, V.V., Nikolaeva, G.V., Dremova, N.N., Knerel'man, E.I., Davydova, G.I., Efimov, O.N., Ionov, S.G., Electroactive polymer coatings on a modified graphite foil as electrodes for supercapacitors. *Prot. Met. Phys. Chem. Surf.*, 55, 321, 2019.
82. Yoon, W.J., Reinhall, P.G., Seibel, E.J., Analysis of electro-active polymer bending: A component in a low cost ultrathin scanning endoscope. *Sens. Actuators A: Phys.*, 133, 506, 2007.

83. Feng, G.H. and Tsai, J.W., Micromachined optical fiber enclosed 4-electrode IPMC actuator with multidirectional control ability for biomedical application. *Biomed. Microdevices*, 13, 169, 2011.
84. Feng, G.H. and Chen, R.H., Fabrication and characterization of arbitrary shaped μIPMC transducers for accurately controlled biomedical applications. *Sens. Actuators A: Phys.*, 143, 34, 2008.
85. Lu, C., Zhao, L., Hu, Y., Chen, W., A molecularregulation strategy towards low-voltage driven, multi degree of freedom IPMC catheters. *Chem. Commun.*, 54, 8733, 2018.
86. Ruiz, S., Mead, B., Palmre, V., Kim, K.J., Yim, W., A cylindrical ionic polymer-metal composite based robotic catheter platform: Modeling, design and control. *Smart Mater. Struct.*, 24, 015007, 2015.
87. Kim, S.J., Pugal, D., Wong, J., Kim, K.J., Yim, W., A bio-inspired multi degree of freedom actuator based on a novel cylindrical ionic polymer-metal composite material. *Rob. Auton. Syst.*, 62, 53, 2014.
88. Eamex Corporation, 2020, http://www.eamex.co.jp/features/koubunshi/ion/iryou.html.
89. Shahinpoor, M. and Kim, K.J., Ionic polymer-metal composites: IV. Industrial and medical applications. *Smart Mater. Struct.*, 14, 197, 2005.
90. Li, H.K., Fan, M., Yue, Y.H., Hu, G.L., He, Q.S., Yu, M., Motion control of capsule-like underwater robot utilizing the swing properties of ionic polymer metal composite actuators. *J. Bionic Eng.*, 17, 281, 2020.
91. Ashton, M.D., Appen, I.C., Firlak, M., Stanhope, N.E., Schmidt, C.E., Eisenstadt, W.R., Hur, B., Hardy, J.G., Wirelessly triggered bioactive molecule delivery from degradable electroactive polymer films. *Polym. Int.*, 70, 467, 2021.
92. Hardy, J.G., Mouser, D.J., Arroyo-Currás, N., Geissler, S., Chow, J.K., Nguy, L., Kim, J.M., Schmidt, C.E., Biodegradable electroactive polymers for electrochemically-triggered drug delivery. *J. Mater. Chem. B*, 2, 6809, 2014.
93. Ralls, A., Kumar, P., Misra, M., Menezes, P.L., Material design and surface engineering for bio-implants. *J. Miner. Met. Mater. Soc.*, 72, 684, 2020.
94. Supronowicz, P.R., Ajayan, P.M., Ullmann, K.R., Arulanandam, B.P., Metzger, D.W., Bizios, R., Novel current conducting composite substrates for exposing osteoblasts to alternating current stimulation. *J. Biomed. Mater. Res.*, 59, 499, 2002.
95. Rivers, T.J., Hudson, T.W., Schmidt, C.E., Synthesis of a novel, biodegradable electrically conducting polymer for biomedical applications. *Adv. Funct. Mater.*, 12, 33, 2002.
96. McLeod, K.J. and Rubin, C.T., The effect of low-frequency electrical fields on osteogenesis. *J. Bone Joint Surg. Am.*, 74, 920, 1992.
97. Shao, S., Zhou, S., Li, L., Li, J., Luo, C., Wang, J., Osteoblast function on electrically conductive electrospun PLA/MWCNTs nanofibers. *Biomaterials*, 32, 2821, 2011.
98. Wu, Y. and Le Gorrec, Y., Optimal actuator location for electroactive polymer actuated endoscope. *IFAC Pap.*, 51, 199, 2018.

99. Puiggalí-Jou, A., Del Valle, L.J., Alemán, C., Drug delivery systems based on intrinsically conducting polymers. *J. Control. Release*, 309, 244, 2019.
100. Palza., H., Zapata, P.A., Angulo-Pineda, C., Electroactive smart polymers for biomedical applications. *Materials*, 12, 277, 2019.
101. Uppalapati, D., Boyd, B.J., Garg, S., Travas-Sejdic, J., Svirskis, D., Conducting polymers with defined micro- or nanostructures for drug delivery. *Biomaterials*, 111, 149, 2016.
102. Krukiewicz, K., Stokfisz, A., Zak, J.K., Two approaches to the model drug immobilization into conjugated polymer matrix. *Mater. Sci. Eng. C*, 54, 176–181, 2015.
103. Ge, D., Tian, X., Qi, R., Huang, S., Mu, J., Hong, S., Ye, S., Zhang, X., Li, D., Shi, W., A polypyrrole-based microchip for controlled drug release. *Electrochim. Acta*, 55, 271, 2009.
104. Ge, J., Neofytou, E., Cahill III, T.J., Beygui, R.E., Zare, R.N., Drug release from electric-field-responsive nanoparticles. *ACS Nano*, 6, 227, 2012.
105. Olvera, D. and Monaghan, M.G., Electroactive material-based biosensors for detection and drug delivery. *Adv. Drug Delivery Rev.*, 70, 396, 2021.
106. Dominika, C.G., Magdalena, S., Adrian, B., Sylwia, G., Wioletta, P., Ewa, Z.G., Sebastian, S., Beata, C., Katarzyna, K., Bifunctional conducting polymer matrices with antibacterial and neuroprotective effects. *Bioelectrochemistry*, 144, 2022.
107. Anna, P.J., Luis, J.V., Carlos, A., Encapsulation and storage of therapeutic fibrin-homing peptides using conducting polymer nanoparticles for programmed release by electrical stimulation. *ACS Biomater. Sci. Eng.*, 6, 2135, 2020.
108. Li, X., Wang, L., Zhang, Z., Kong, D., Ao, X., Xiao, X., Electroactive high-temperature shape memory polymers with high recovery stress induced by ground carbon fibers. *Macromol. Chem. Phys.*, 220, 1900164, 2019.
109. Zheng, N., Fang, Z., Zou, W., Zhao, Q., Xie, T., Thermoset shape-memory polyurethane with intrinsic plasticity enabled by transcarbamoylation. *Angew. Chem.*, 128, 11593, 2016.
110. Basit, A., L'Hostis, G., Durand, B., The recovery properties under load of a shape memory polymer composite material. *Materialwiss. Werkstofftech.*, 50, 1555, 2019.
111. Peng, B., Yang, Y., Gu, K., Amis, E.J., Cavicchi, K.A., Digital light processing 3D printing of triple shape memory polymer for sequential shape shifting. *ACS Mater. Lett.*, 1, 410, 2019.
112. Xie, Y., *Applications of bistable electroactive polymers as rewritable photonic paper, smart windows and wearable pressure sensors*, Ph.D. Thesis, University of California, Los Angeles, CA, USA, 2019.
113. Koo, I.M., Jung, K., Koo, J.C., Nam, J.D., Lee, Y.K., Choi, H.R., Development of soft-actuator-based wearable tactile display. *IEEE Trans. Robot.*, 24, 549, 2008.

114. T. Rider, S. Taite, Intel Corp., Wearable adaptive electroactive polymer fabric. United States Patent US 10,334,899, 2 July 2019.
115. Forouharshad, M., King, S.G., Buxton, W., Kunovski, P., Stolojan, V., Textile-compatible, electroactive polyvinylidene fluoride electrospun mats for energy harvesting. *Macromol. Chem. Phys.*, 220, 1900364, 2019.
116. Bhattacharya, S., Khan, S., Sil, T., Bepari, B., Bhaumik, S., IPMC based data glove for ginger motion capturing. *Conference on Advances in Robotics*, 1, 2015.

6

Bioinspired Hydrogels Through 3D Bioprinting

Farnaz Niknam[1], Vahid Rahmanian[2], Seyyed Mojtaba Mousavi[3*], Seyyed Alireza Hashemi[4], Aziz Babapoor[1] and Chin Wei Lai[5]

[1]*Department of Chemical Engineering, University of Mohaghegh Ardabil, Ardabil, Iran*
[2]*Centre of Molecular and Macromolecular Studies, Polish Academy of Sciences, Sienkiewicza, Lodz, Poland*
[3]*Department of Chemical Engineering, National Taiwan University of Science and Technology, Taipei, Taiwan*
[4]*Nanomaterials and Polymer Nanocomposites Laboratory, School of Engineering, University of British Columbia, Kelowna, BC, Canada*
[5]*Nanotechnology and Catalysis Research Centre (NANOCAT), Institute for Advanced Studies (IAS), Universiti Malaya (UM), Kuala Lumpur, Malaysia*

Abstract
Hydrogels have been considered in scientific and industrial fields due to their unique properties and have been produced and studied for use in different applications. In recent years, three-dimensional printing technology (3D) has attracted much interest from researchers, as it is used in many industrial sectors to rapidly and easily create complex structures and materials. It offers many advantages over conventional production techniques, such as using new materials and processes not available in conventional production techniques. It is common to use hydrogen-based bioinks for the three-dimensional printing of biological tissues. This chapter will discuss three aspects of 3D bioprinting: hydrogel inks as inks for 3D bioprinting and polymers used to create bioinspired hydrogels.

***Keywords*:** Medicine, hydrogel, 3D bioprinting

**Corresponding author*: kempo.smm@gmail.com

Inamuddin, Tariq Altalhi and Ashjan Alrogi (eds.) Biomimicry Materials and Applications, (147–168)

6.1 Introduction

Hydrogels are three-dimensional networks of interconnecting polymer chains which can be chemically or physically crosslinked and which are capable of absorbing water and aqueous solutions. Amorphous gels are often called hydrogels. However, they are not soluble in water at any concentration [1–3]. Covalent bonding, physical crosslinking (such as molecular entanglements), and secondary forces may all generate a hydrogel (such as ionic bonding, hydrogen bonding, and hydrophobic forces) [3]. Hydrogels obtained by chemical cross-linking (covalently attached hydrogels) are more stable and adjustable than those obtained by physical cross-linking. The cross-linking process involves cross-linking between complementary groups such as aldehydes, radical polymerization, activation of enzymes, and condensation of carboxylic acids with hydroxyl or amine groups [4]. Due to the chemistry, function, and structure of hydrogels, they may be utilized in a variety of applications such as, shape memory, tissue engineering, weak stimuli, shape morphing, sensors, soft robotics, drug delivery and artificial muscles Figure 6.1 [5, 6]. Generally, these materials can be classified into two categories: Normal and intelligent [7, 8]. The smart hydrogel can reply to outside stimuli, like

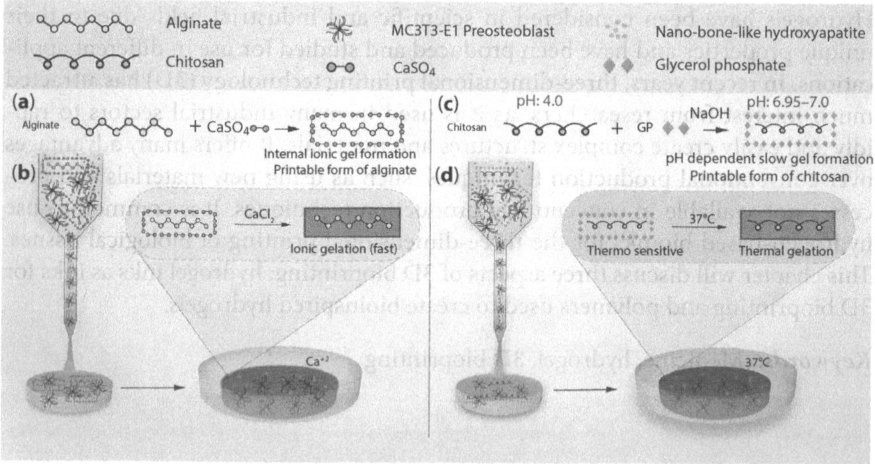

Figure 6.1 Schematic illustrations demonstrate the use of alginate and chitosan hydrogels as bioinks in 3D bioprinting for bone tissue engineering [12] (reprinted from ref. 12 with permission from Elsevier [open access]).

temperature and antigen, and change its shape and volume [8]. Hydrogels are biologically beneficial to meet different needs [6]. Due to their potential and prospective uses in various disciplines, such as prosthesis, temperature-sensitive materials have lately attracted a lot of interest [7]. Biomedical researchers have considered bioinspired hydrogels due to their fibrous architecture. Traditional hydrogels lack fibrous structures similar to those present in tissues and have poor mechanical properties [9]. Their remarkable and unique properties, which are also highly adjustable, make it possible for materials scientists to incorporate bioinspired elements into hydrogels [10]. Researchers have also begun to investigate the chemical and physical characteristics of hydrogel which can be 3D printed. In addition to its tensile strength, swelling behavior, hardness, and electric conductivity, this material also has a number of other properties. Three-dimensional printing produces hydrogels that are designed to interact with each other, including dynamic covalent bond interactions, hydrogen bond interactions, host and super molecular guest interactions, and hydrophobic interactions, 3D printing hydrogels are robust and have a high hardness that compensates for the fragility which exists in most traditional hydrogels and smart hydrogels and leads to their greater use in the field of materials science. Just as intelligent hydrogels are stimulus-responsive, so are 3D-responsive stimulus-responsive hydrogels; for example: temperature-responsive 3D hydrogel, moisture-responsive 3D hydrogel, photo-responsive 3D hydrogel, dual-response photothermal 3D hydrogel and temperature, 3D pressure, magnetically responsive 3D hydrogel, etc. [11]. In order to enable new functions, Bio/Inspiration, introduces new structures. Hydrogels have inherent physical, chemical, and biological properties providing a platform for new/improved functions. Numerous types of hydrogel-based biomimetics draw inspiration from various sources; for example, Water-soluble hydrogels made from fibrin or collagen are biodegradable and non-toxic, with the capability of mimicking ECM mechanics via the self-forming, three-dimensional fibers which can be injected locally. Furthermore, when stimulated by integrin ligands, they enhance the angiogenesis of the tissues. This function can trigger cellular signaling, binding, proliferation, and secretion. Besides having antibacterial properties, these gels may also adhere to surfaces without substantial displacement during movement. Therefore, they are an excellent choice for use in tissue repair devices compared to traditional adhesives [10].

There are two topics presented in this chapter: 3D bioprinting and hydrogel inks for 3D bioprinting.

6.2 Bioinspiration

As a general category, "bioinspired" and "biomimetic" denote anything inspired by or imitates biological processes. At the same time, they are adapted from nature and are trying to find new solutions in various sciences for unsolved problems for the production of biological materials [10]. According to research on bioinspired materials, natural materials are often composed of minerals and biological polymers, which may offer higher performance than synthetic equivalents. Even while nature increases excellent structural patterns for the performance of the materials and their gradational properties exceed the potential of traditional production methods. Therefore, we might seek new construction techniques and create new approaches. Engineers can optimize designs to regulate the mechanical properties of materials and build complex structures inspired by natural architecture; they use additive fabrication sophisticated biological and chemical synthesis approaches using 3D printing techniques to print synthetic materials to produce new inks similar in nature to the components. 3D printing of buildings inspired by nature may be used to overcome the restrictions of renewable energy, engineering, medicine, and many other sectors Figure 6.2 [13].

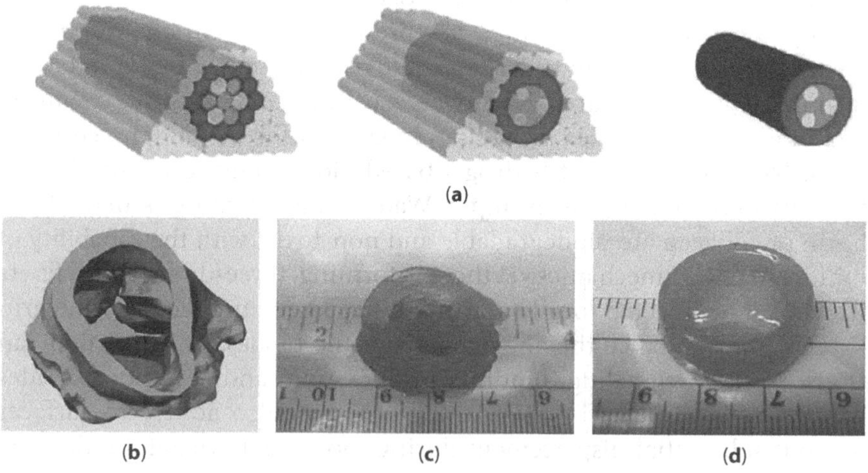

Figure 6.2 Hollow tissue produced through bioprinting. (a) Multiple luminaires on a nerve graft. (b–d) Heart valves: (b, c) Printed micro-tomography (μCT) model of the aortic valve, images of the aortic valve duct. (d) Aortic valve by encapsulating human aortic valve (HAVIC) interstitial cells within the leaflets [14] (reprinted from ref. 14 with permission from Elsevier (Open access)).

6.3 3D Bioprinting

Three-dimensional printing has attracted much attention in recent years. It is increasingly used in various fields to quickly produce intricate structures and materials beyond the limitations of conventional production methods [11, 15, 16]. Bioinspired materials with complex compositions and structures can be made more efficiently using 3D printing with multi-material, multiscale capabilities (Table 6.1) [17]. It has applications in agriculture, education, energy, industry, aerospace, healthcare, transportation, communications, culture, military affairs, national defense, etc. (Figure 6.3) [18]. Some applications of bioinspired structures processed by 3D printing include stable vascular microfibers, biochemical materials, such as lotus root for cell delivery and reconstruction, branch honeycomb structures for cells immigration research cancer, and other programs [19]. Typically, printing techniques include high temperatures, chemical solvents, and crosslinking agents incompatible with live cells and biological materials. Therefore, utilizing acceptable printing materials with excellent printability, mechanical characteristics, desired degradation, and biocompatibility for tissue architectures is an important problem in 3D bioprinting. Natural or synthetic biomaterials are commonly used for 3D bioprinting. In natural derivatives (like collagen, fibrin, gelatin, hyaluronic acid, chitosan, and alginate), the main advantage is their inherent bioactivity. They are typically similar to extracellular matrix (ECM) and outstanding biocompatibility. Nevertheless, natural biomaterials are not good in mechanical properties even if they are crosslinked, so materials with more significant advantages than natural materials can be used, like synthetic materials produced with specific physical virtues. Thus, advanced stimulus-responsive biomaterials can be developed for three-dimensional bioprinting of synthetic biological structures which can adapt to a complex and dynamic physiological environment [18]; for a program, according to the materials used and also how to glue the layers of the final product, the type of 3D printer is selected [20]. Several approaches, such as bioprinting techniques including inkjet, laser-assisted, and extrusion and stereolithography, are used in 3D bioprinting to create bioinspired hydrogels [4, 19, 21].

6.3.1 Inkjet Bioprinting

It is a contactless reprogramming method that reproduces digital data on a layer by receiving it from a computer [43]. Improved versions of

Table 6.1 A short summary of recent bioprinting studies.

Bioprinting methods	Bioink used	Applications	Ref.
Inkjet	Fibrin, collagen and thrombin	Scaffolding for the delivery of stem cell therapy	[22]
Inkjet	(PEGDMA, PEG) with gelatin methacrylate (GelMA)	Suitable for cartilage and bone tissue engineering	[23]
Laser-assisted	Sodium alginate	Tissue engineering	[24]
Extrusion-based bioprinting, laser-based bioprinting (*e.g.* stereolithography)	Chitosan	Bone tissue, vascular tissue, cartilage tissue (*e.g.*, human ear), skin tissue	[25–30]
Extrude	SF-Collagen	Knee cartilage; tissue engineering	[31]
Extrusion-based bioprinting	Agarose	Vascular tissue	[32, 33]
Inject	SF-Gelatin	Artificial Implant/ Cartilage tissue	[34–36]
Laser-assisted	Matriderm®	3D skin for wound regeneration	[37]
Extrusion-based bioprinting deposition, laser-based bioprinting droplet-based bioprinting	Alginate	Bone, vascular tissue, cartilage	[38–40]
Inject	Gelatin-methacrylamide/ HA	Fabrication of tissue-engineered cartilage constructs	[41]
Laser-assisted	HMSCs	Could create complex tissue structure	[42]

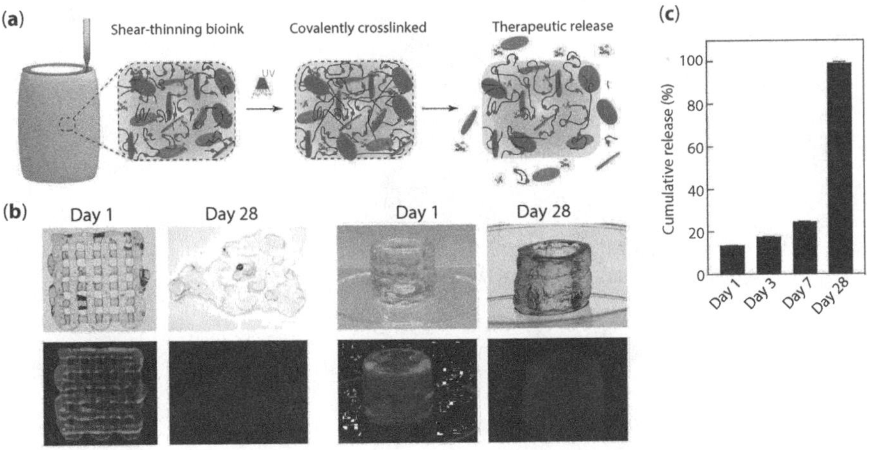

Figure 6.3 The production and dissemination of three-dimensional medicinal materials. (a) Charged nanosilicates with a significant surface area may segregate therapeutic proteins inside a 3D-printed structure. When printing print networks, the therapy will be published. (b) 3D-printed structure releases fluorescent-tagged proteins. (c) Release of medicines over time [12] (reprinted from ref. 12 with permission from Elsevier (Open access)).

commercial 2D inkjet printers were the first inkjet printers used in bio-printing applications. An inkjet printer uses sound or heat forces to spray droplets onto a substrate. These ink droplets can then be used to support or construct the final structure. Numerous experimental findings have indicated that the benefits of inkjet bioprinting are fast print speed since the print head may support parallel operation mode; inexpensive cost owing to the comparable structure of commercial printers; and a reasonably good survival rate (typically 80% to 90%). The risks of cells and materials exposed to temperature and mechanical stressors, frequent nozzle obstruction, the downward direction of droplets, unsatisfactory encapsulation of cells, and irregular droplet size have resulted in substantial limits in tissue engineering applications [21]. Three different inkjet printing methods can be distinguished: continuous inkjet printing, drip inkjet printing, and electro-hydrodynamic jet printing. Inkjet bioprinting is the largest and most popular category of bioprinting (Figure 6.4) and can be achieved in thermal, piezoelectric, or electrostatic [44]. Compan *et al.* [45] report that silk fibroin functions as a natural protein and can also be used for tissue engineering. Our study presents a novel two-step gel fabrication process for silk fibroin printing.

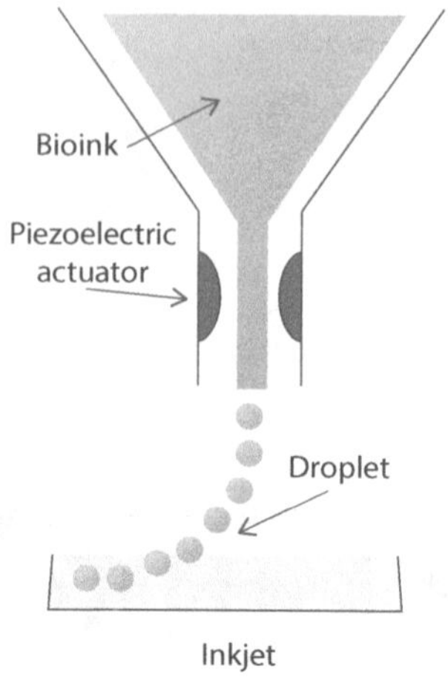

Figure 6.4 Schematic for the inkjet bioprinting method [47] (reprinted from ref. 47 with permission from mdpi [open access]).

As part of the inkjet process, the sacrificial hydrogel is composed of alginate; to promote rapid gel formation, alginate and silk fibrin is used to produce calcium alginate. Using a two-step gel manufacturing procedure, cell-filled fibrin structures can be bioprinted in three dimensions and cultivated for an extended period. After calcium chelation, liquid alginate is produced, which retains the structure. In this way, silk fibroin hydrogels are produced. The proliferation of NIH 3T3 fibroblasts occurs with the dispersion of the fibroblasts through hydrogels following printing. Silk fibroin can be printed in three-dimensional form using the proposed two-step gel method.

6.3.2 Extrusion Printing

Researchers have recently used extrusion-based methods for fabricating scaffolding as an alternative method [44]. Over the last several years, tissue engineers and bio fraction fabricators have been more interested in extrusion-based bioprinting, also known as direct writing [46]. 3D printing uses an extrusion-based process, which uses a moveable nozzle operating as the extrusion printing head in direct ink writing mode to deposit layers by layers of molten or semi-molten polymers, polymer solutions, pastes, or dispersions. Figure 6.5 is an illustration of how the different types of ink leakage manifest themselves in extrusion-based 3D printing [43]. Three-dimensional cell carriers can be manufactured using extrusion bioprinting, a technique for regenerating tissue. Pre-polymer

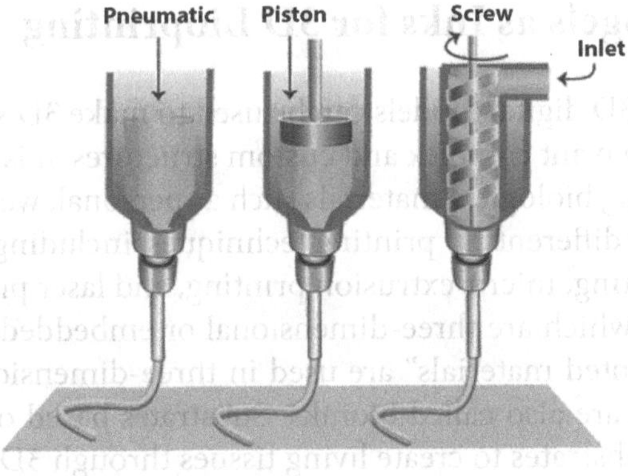

Figure 6.5 A diagram illustrating the operating principles of extrusion-based 3D printing [43] (reprinted from ref. 43 with permission from Elsevier [open access]).

solutions get ready first, and there is the ability to print nearly all kinds of high-density granular pre-polymer solutions and different viscosities with extruded bioprinters. Unlike small droplet printing on a platform, extrusion bioprinting continuously places hydrogel filaments 150 to 300 microns in diameter to produce three-dimensional structures. Common methods of extrusion bioprinting include pneumatic, piston, and screw distribution.

Pneumatic, screw, and piston distribution are typical extrusion bioprinting techniques. As the pneumatic distribution relies on air pressure for its driving power, the piston distribution and pneumatic distribution, on the other hand, use rotating and vertical mechanical forces to initiate printing. The capacity of a bioprinter to extrude biodegradable materials is contingent upon three factors: viscosity adjustment, pre-extrusion bio injection phase, and material-specific bioproduction window. The use of extrusion bioprinters has enabled the production of branched vascular trees and *in vitro* drug delivery. However, it may take a considerable amount of time to construct complicated high-resolution structures. The microfluidic chamber's structures are microfabricated from therapeutically relevant tissue size. The composition of cells by bioactive agents is also comfortable because there is no heating process involved. Extrusion bioprinters, as opposed to inkjet 3D bioprinters, can produce a continuous flow of biomaterials, resulting in ease of use and a wider range of biomaterials. As a result, they have a wider range of biomaterials to choose from, such as cellular matrices, polymers, spheres, cellular hydrogels, and granules, and they are easier to work with [21, 47].

6.4 Hydrogels as Inks for 3D Bioprinting

3D printing of 3D digital models can be used to make 3D structures. Since 3D printing can print complex and custom structures, it is a versatile technique for making biological materials such as personal, wearable implants. There are many different 3D printing techniques, including stereolithography, inkjet printing, micro-extrusion printing, and laser printing [4]. Cells or cell masses, which are three-dimensional or embedded in biomaterials and are "bioprinted materials" are used in three-dimensional bioprinting techniques and are also called bioinks. Substrates based on hydrogels are often used in substrates to create living tissues through 3D printing. These bioinks are composed of gel-based materials combining biological agents and have their own physical and chemical properties due to the synthesis method and components [16]. In 3D printing, the inks we use must be able to flow out and solidify quickly [48]. In 3D bioprinting, it is necessary to

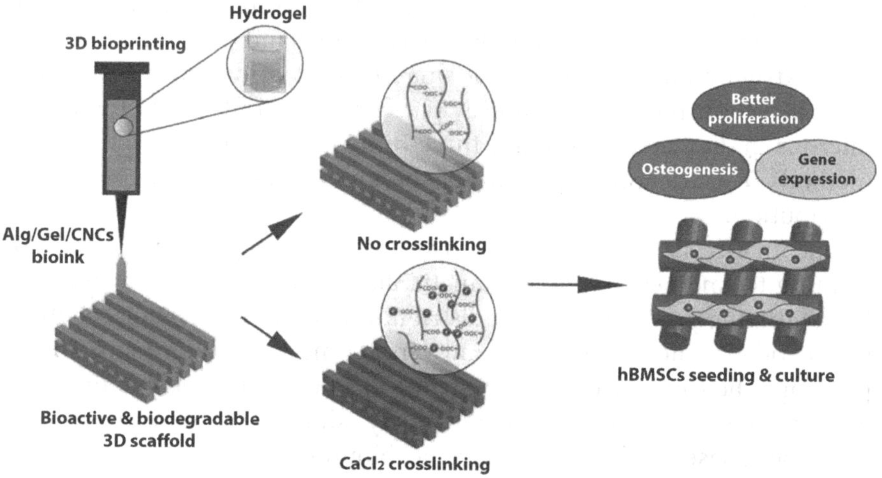

Figure 6.6 Schematic representation of 3D bioprinting using Alg/Gel/CNCs hydrogels for tissue engineering. Alg; Alginate; Gel: Gelatin; CNCs: Cellulose nanocrystals [49] Wei *et al.* [50] synthesized bioinspired Ca-PAA-SA-CNT hydrogels with self-healing, elastic, and 3D printing capabilities in a simple manner that have ionic and electron conductivity. By integrating a layer of dielectric polyethylene elastomer with two printed hydrogels in the shape of a sandwich, with copper electrodes spanning the space between the layers, an integrated strain sensor capable of detecting changes in capacity and strength was built. This pressure sensor detects the signals inscribed on it and the delicate motions in the human body, such as bending the finger, bending the knee, and breathing, in a sensitive, precise, and continuous manner. We anticipate this conductive hydrogel in skin-wearable devices, such as bioinspired devices, since it combines all the aforementioned capabilities (reprinted from ref. 49 with permission from Elsevier (License Number: 5347161023974)).

find and select the appropriate biological ink and pay attention to this issue because it supports cell growth and maturation by preparing specific tissue microenvironments [16].

Previous research has shown that hydrogels may be created for use as inks in 3D printing with synthetic polymers, such as PF 127, polyethylene oxide/polyglycol, and PVA. In another study, PEtOzi-based bioink was used in scaffold printing. PEtOzi hydrogel has reversible properties and is related to click chemistry, and also because this hydrogel has functional groups which can be easily changed. If these hydrogels are used as biowinkers to make biological materials, it should be noted that the tissue to be replaced must have mechanical properties close to that tissue, which most conventional hydrogels do not have good mechanical virtues. Therefore, modifications like IPN virtues, nanocomposites, and block copolymers should be performed on bioinks produced from hydrogels (Figure 6.6) [48].

6.5 Polymers Used for Bioinspired Hydrogels

Among the bioprinting materials used to make physiologically complex human heterogeneous tissues are biomaterials like synthetic hydrogels (polyurethane, polyethylene glycol) and natural hydrogels like collagen, fiber, silk, hyaluronic acid, chitosan, cellulose. Have been studied. It is also possible that some decellularized extracellular matrix (dECM) hydrogels are a mix of diverse proteins in natural tissue or organ ECM and are termed native hydrogels, comparable to bioinks [16]. These materials can be used in hydrogels used for 3D bioprinting, which improves mechanical properties; by chemical modification on natural polymers, their ability to print as bioink can be increased, which is very important (Figure 6.7) [51–53].

6.5.1 Alginate

Alginate, or alginic acid, is an anionic polysaccharide, Cold water dissolves this polymer [19]. The physicochemical characteristics of the alginate solution are directly connected to the ratio between M and G blocks. A higher G/M ratio improves polymer structure and mechanical characteristics, whereas a lower G/M ratio improves flexibility [16]. Obtained from marine plants and microbes, it is produced by brown seaweed and soil bacteria. Hence it is present in nature. One of the most prevalent types of alginates used in the market is sodium alginate. The processes used in acidic or alkaline purification to produce sodium alginate from brown seaweed

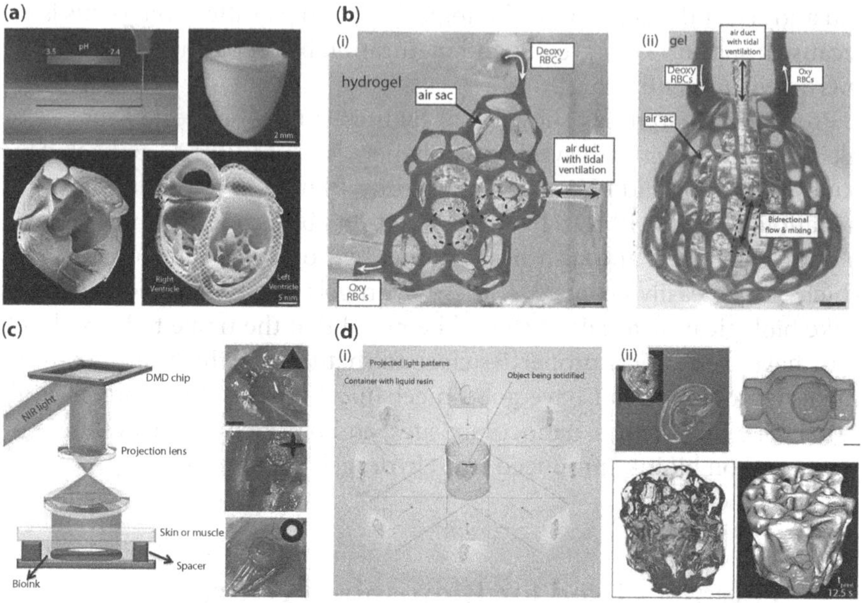

Figure 6.7 Representation advances in 3D bioprotection activation techniques: (a) FRESH technique: 3D collagen bioprinting for reconstruction of human heart components. (b) The "SLATE Technique" refers to complex 3D structures, which may alveolate and are printed using a digital light processing method. (i) Image of a hydrogel printed during perfusion of red blood cells (RBCs), (ii) this image depicts the distal lung component of a printed hydrogel when it is being perfused with blood cells and oxygenated. (c) Noninvasive bioprinting *in vivo*: near-infrared (NIR) light polymerization of a three-dimensional bioink of a human ear-shaped structure. (d) Bioprinting using volumetric stereolithography involves baking a hydrogel precursor in a chamber that rotates while exposing it to different optical patterns. This results in the production of complex tissue models. (i) diagram showing the three-dimensional bioprinting of human ear models using tomographic projections; (ii) Complex structures printed using volumetric printing [54] (reprinted from ref. 54 with permission from Elsevier [open access]).

are relatively simple. Processing problems are mainly due to the separation of sodium alginate from viscous residues. Alginate provides a number of industrial benefits, including cheap cost, strong biocompatibility, and a wide range of functional groups. Slowly absorbs 200 to 300 times its weight in water, becoming a sticky gum [55]. Alginate and nanocellulose may be combined to create a biomaterial which can be tested for biocompatibility with human nasal cartilage cells. This composite roll material may be manufactured using an extrusion-based 3D printer with optimum process parameters [19]. Alginates are used in various industries as viscosities, stabilizers, gels, filmmakers, and water-binding agents. Heat-resistant gels are

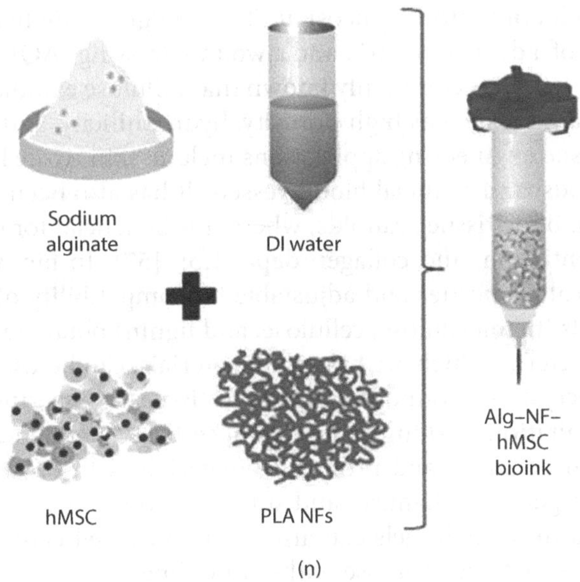

Figure 6.8 Composite materials: one example of this would be the incorporation of polylactic acid nanofibers (NF) into cell-filled alginate hydrogels (Alg) in order to stimulate human mesenchymal stem cell proliferation and cartilage development (hMSCs) [14] (reprinted from ref. 14 with permission from Elsevier [open access]).

formed using this polymer. In the biological manufacturing process, it is used as a hydrogel [55]. Alginate hydrogels and alginate composites are often employed as cellular "bioinks" in several 3D bioprinting methods [56] because to strong biocompatibility (such as non-immunogenicity, low toxicity), quick biodegradability, and chemical gelling capabilities (e.g., allogeneic characteristics) (Figure 6.8).

6.5.2 Cellulose

One of the most common polysaccharides found in nature is cellulose [57]. It is made up of D-glucose units in a straight line. D-glucose units are condensed with beta glycoside linkages to form cellulose (1→4). As a structural component of their primary cell walls, chlorophyll is present in green plants, various kinds of seeds, and algae. Despite being a hydrophilic linear polymer, it is impervious to water and organic solvents because of its strong hydrogen bonds between polymer chains. Numerous cellulose derivatives, including methylcellulose, hydroxypropyl cellulose, hydroxypropyl methylcellulose, and carboxymethylcellulose, may be produced by converting cellulose into D-glucose units using glycoside hydrolase. The effectiveness of a delivery matrix and a wound dressing (AQUACEL®) has been studied [60]. It is commonly known that cellulose exhibits mesh very preferred qualities, such as high porosity, hydrophilicity, and mechanical strength. Tissue engineering applications include skin tissue healing, biological scaffolds, and artificial blood vessels. It has also been investigated extensively in bone tissue scaffolds, where it is beneficial for cell proliferation, differentiation, and collagen deposition [57]. In bio-3D printing, the mechanical properties and adjustable biocompatibility of lignocellulosic materials (hemicellulose, cellulose, and lignin) obtained from fungi, wood, and bacteria, which are biostable materials, can be used as nanofibers or nanocrystals. Wound dressing, and tissue engineering are widely used as an imitation of skin tissue. Cellulose-based nanofibers (CNF) in hydrogels can be stable and promising materials with outstanding biocompatibility, good mechanical, and sufficient compressive strength and elasticity, and these hydrogels can also be used as well elastic hydrogels. 3D bioprinters that use cellulose without holding structures may biofabricate inexpensive scaffolds for patients who can simulate various tissue types [19]. Schmidt *et al.* [58] As a drug carrier for triamcinolone acetonide, nanostructured mesoporous silica, SBA-15, was used; and it was discovered that the drug in SBA-15 pores lowered its pore area and volume, while enhancing the apparent water solubility of triamcinolone acetonide. The hydrophilic polymer film was then adsorbed on a hydrophilic ink made from carboxymethylcellulose containing SBA-15 containing the drug employing a semi-solid extrusion (SSE) technique. These three-dimensional-printed films showed complete release of the drug, and their mucosal adhesion was improved *in vitro* with the presence of SBA-15 containing triamcinolone acetone, and it can be concluded that it is promising in the treatment of oral mucosa. This is the first report to propose the

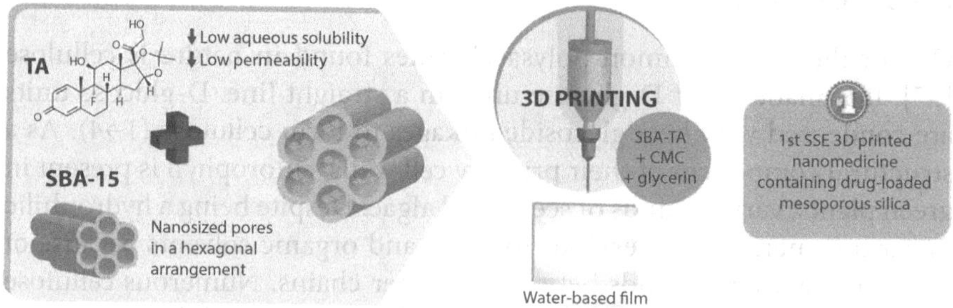

Figure 6.9 Mesoporous silica loaded with drugs on carboxymethyl cellulose hydrogel (reprinted from ref. 58 with permission from Elsevier [license number: 5347160406042]).

development of SSE three-dimensional printing nanoparticles containing drug-containing mesoporous silica and an innovative platform for the development of water-based mucosal adhesive formulations containing a hydrophobic drug (Figure 6.9) [58].

6.5.3 Chitosan

Chitin may be broken down into chitosan by chemical synthesis. It is the natural biopolymer with the second-highest abundance, and it is a cationic polysaccharide. It is non-toxic, biocompatible, inexpensive to extract, biodegradable, and readily available, and it has a plentiful supply. These are some of its exceptional qualities [59–62]. Therefore, it has different applications (biomedical, food, cosmetics, pharmaceutical, etc. industries) [63]. Because chitosan is similar in structure and composition to glycosaminoglycans (GAGs), it is biocompatible and non-toxic [64]. Chitosan remains insoluble in neutral and alkaline states [63]. In chitosan molecules, both groups - NH2 and -OH- cause chemical changes in their structure and create different desired properties [6]. Physical characteristics of inks chitosan-based biointroduction of different enhancer additives enable the printing of chitosan-based bioinks. Additionally, chitosan-based hydrogels exhibit significant bioactivity and superior cell survival. Chitosan-based bioinks are prospective candidates for 3D bioprinting applications due to the combination of these customizable and flexible characteristics (Figure 6.10) [65]. In Long *et al.* [66], the feasibility of a 3D-printed CS-PEC hydrogel containing the local anesthetic lidocaine hydrochloride (LDC) was proven for the first time, indicating that a 3D-printed hydrogel might be a promising choice for wound dressing.

Zhou *et al.* [67] described a novel chitosan ink which may be used for manufacturing chitosan hydrogels with increased strength. The suggested printing process creates biocompatible scaffolds. In addition, the wound healing model demonstrates another way that printed chitosan scaffolds are appropriate for use in tissue engineering.

6.5.4 Fibrin

Fibrin is a protein with a fibrous structure produced from fibrinogen in the presence of thrombin protease. Fibrin is derived from blood and is synthesized by the polymerization of fibrinogen. Fibrin is an example of a protein, which is not spherical. They are often used as hemostatic agents, sealants, and surgical adhesives in clinics. Fibrin can be gel under physiological

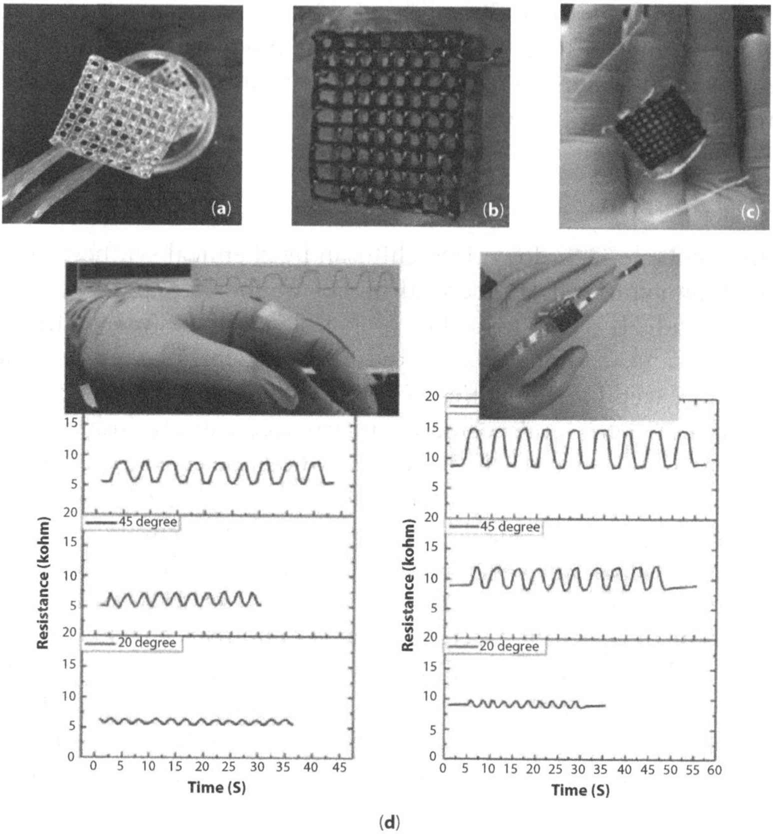

Figure 6.10 Three-dimensional-printed structures of conductive and non-conductive bioinks. (a) A standard extrusion bioprinting grid structure made out of chitosan, and (b–d) a conductive 3D printing sensor made out of chitosan and acrylic acid and encapsulated in PDMS. Displayed as a sensor in the form of a bar (on the left) and as a 3D-printed mesh (on the right) is the resistance response of the hydrogel test subjected to varying degrees of bending (right) [44] (reprinted from ref. 44 with permission from Elsevier [open access]).

conditions without modifying the main components. Other natural hydrogels may be needed to create stable gels that require chemical modification or harsh crosslinking conditions. Because fibrinogen has a low viscosity, it is only occasionally used to print three-dimensional structures. However, fibrin bioprinting can be achieved by combining higher viscosity materials to produce composite structures. In these methods, there is a significant requirement for crosslinking other components (such as alginate and gelatin) to produce three-dimensional structures. Because fibrin can quickly form a stable gel, it cannot be said that this additional crosslinking should be used, so the inclusion of materials, such as alginate, usually improves the printing and mechanical properties of the three-dimensional structure without regard to biological effect. Depending on the crosslinking conditions, the mechanical properties of fibrin can vary, which can allow the phenotypic behavior of the fusion cells to be precisely controlled. There is an important challenge because there are mechanical problems in making protein hydrogel scaffolds with a specific microstructure; for example, 3D printing of extrusion-based fibrin is generally impossible because it has a low viscosity due to the precursor fibrinogen solution. It is used in various fields of biomedicine, such as wound healing and tissue repair. Fibrin hydrogels are ideal for encapsulation, transport, and cell culture because they are extremely compatible with cell viability. As a result of proteolytic enzymes, fibrin may decompose quickly. Fibrin can also be used as a biological substance, as it can facilitate communication between cells due to its nonlinear elasticity [16, 68]. In a study by Sriphutkiat *et al.* [68], fibrinogen was cultured as a support structure for adhesion and elongation. GelMA and fibrinogen were coupled to boost the printing capability of bioinks, as the GelMA-fibrin double crosslinking capacity was found to give a more robust and stable cell-filled structure. Creating a crosslinked polymer network may lead to an increase in stability. Abelseth *et al.* [69] present new methods for preparing fibrin-based bioinks and printing human tissues using the RX1 Aspect Biosystems printer. Our new Bioink and Aspect Biosystems "RX1" printer makes it possible to create high-power 3D human textures together.

6.5.5 Silk

Silk is an insoluble fibrous protein; silk chains form alternating hydrophilic and hydrophobic areas similar to block polymers, which may provide amphiphilic characteristics and build semi-crystalline structures via cross-linking hydrophobic interactions in the material. As a result of the hierarchical self-organization of silk, several different processing techniques have resulted in the creation of silk in numerous shapes (including fibers, solids, hydrogels, yarns, and stitches, among others) that have a combination of optimal strength, elasticity, and anti-allergenic properties. Biodegradability, biocompatibility, medical applications, and mechanical qualities have all been the subject of substantial research. The inkjet and extrusion printing processes are the primary methods used for 3D-producing silk. Because of 3D printing, silk can now be produced in a wider variety of patterns, which expands its potential uses. Silk is commonly used to provide better viscosity and mechanical properties for printing, and this protein can be combined with natural and synthetic polymers in biotin. Silk bioink has been used to make medical devices and tissues. Hydrogels made from silk are ideal for regenerative medicine and tissue engineering, making them an interesting choice for the development of tissue transplants. Silk scaffolds offer a variety of uses, including tissue bioengineering, wound healing, bone regeneration and cartilage repair, drug release, and three-dimensional bioprinting of biological structures [16, 52, 70].

6.6 Conclusion

Due to the unique properties of hydrogels, scientists can incorporate bio-inspired elements into hydrogels. In recent years, 3D bioprinting has also received much attention, inspired by nature to influence the 3D printing of structures. Inspired by nature to influence the 3D printing of structures, it helps us overcome the limitations of medicine, renewable energy, engineering, and many other fields. The method of making bioinspired hydrogels with 3D bioprinting has changed, and applications of bioinspired hydrogels have led to the expansion of applications such as pharmaceutical, tissue regeneration, and dressing.

References

1. Srivastava, A. *et al.*, Polymers in drug delivery. *J. Biosci. Med.*, 4, 1, 69–84, 2015.
2. Torres-Luna, C. *et al.*, Hydrogel-based ocular drug delivery systems for hydrophobic drugs. *Eur. J. Pharm. Sci.*, 154, 105503, 2020.
3. Shi, Y. *et al.*, Energy gels: A bio-inspired material platform for advanced energy applications. *Nano Today*, 11, 6, 738–762, 2016.
4. Unagolla, J.M. and Jayasuriya, A.C., Hydrogel-based 3D bioprinting: A comprehensive review on cell-laden hydrogels, bioink formulations, and future perspectives. *Appl. Mater. Today*, 18, 100479, 2020.
5. Qiu, Y. and Park, K., Environment-sensitive hydrogels for drug delivery. *Adv. Drug Delivery Rev.*, 53, 3, 321–339, 2001.
6. Fu, J., Hydrogel properties and applications. *J. Mater. Chem. B*, 7, 10, 1523–1525, 2019.
7. Zhang, X., Wu, D., Chu, C.-C., Synthesis and characterization of partially biodegradable, temperature and pH sensitive Dex–MA/PNIPAAm hydrogels. *Biomaterials*, 25, 19, 4719–4730, 2004.

8. Zhang, S.B. *et al.*, Poly (N-isopropylacrylamide)-based comb-type grafted hydrogel with rapid response to blood glucose concentration change at physiological temperature. *Polym. Adv. Technol.*, 19, 8, 937–943, 2008.
9. Chen, Y. *et al.*, Bio-inspired hydrogels with fibrous structure: A review on design and biomedical applications. *Biomater. Adv.*, 136, 212799, 2022.
10. Seidi, K. *et al.*, Bioinspired hydrogels build a bridge from bench to bedside. *Nano Today*, 39, 101157, 2021.
11. Liu, C. *et al.*, Hydrogel prepared by 3D printing technology and its applications in the medical field. *Colloid Interface Sci. Commun.*, 44, 100498, 2021.
12. Heid, S. and Boccaccini, A.R., Advancing bioinks for 3D bioprinting using reactive fillers: A review. *Acta Biomater.*, 113, 1–22, 2020.
13. Gu, G.X. *et al.*, *MIT industrial liaison program*, 2015.
14. Zhang, B. *et al.*, 3D bioprinting: A novel avenue for manufacturing tissues and organs. *Engineering*, 5, 4, 777–794, 2019.
15. Papaioannou, T.G. *et al.*, 3D bioprinting methods and techniques: Applications on artificial blood vessel fabrication. *Acta Cardiol. Sin.*, 35, 3, 284, 2019.
16. Fatimi, A. *et al.*, Natural hydrogel-based bio-inks for 3D bioprinting in tissue engineering: A review. *Gels*, 8, 3, 179, 2022.
17. Li, T. *et al.*, 3D printing of bioinspired biomaterials for tissue regeneration. *Adv. Healthcare Mater.*, 9, 23, 2000208, 2020.
18. Mao, H. *et al.*, Recent advances and challenges in materials for 3D bioprinting. *Prog. Nat. Sci.: Mater. Int.*, 30, 5, 618–634, 2020.
19. Whenish, R. *et al.*, A framework for the sustainability implications of 3D bioprinting through nature-inspired materials and structures. *Bio-Des. Manuf.*, 5, 2, 412–423, 2022.
20. Choonara, Y.E. *et al.*, 3D-printing and the effect on medical costs: A new era? *Expert Rev. Pharmacoeconomics Outcomes Res.*, 16, 1, 23–32, 2016.
21. Nie, L., Wang, C., Deng, Y., Shavandi, A., Bio-inspired hydrogels via 3D bioprinting. *Biomimetics*, 10, 2020.
22. Skardal, A. *et al.*, Bioprinted amniotic fluid-derived stem cells accelerate healing of large skin wounds. *Stem Cells Transl. Med.*, 1, 11, 792–802, 2012.
23. Gao, G. and Cui, X., Three-dimensional bioprinting in tissue engineering and regenerative medicine. *Biotechnol. Lett.*, 38, 2, 203–211, 2016.
24. Guillotin, B. *et al.*, Laser assisted bioprinting of engineered tissue with high cell density and microscale organization. *Biomaterials*, 31, 28, 7250–7256, 2010.
25. Liu, C. *et al.*, Bioprinted chitosan and hydroxyapatite micro-channels structures scaffold for vascularization of bone regeneration. *J. Biomater. Tissue Eng.*, 7, 1, 28–34, 2017.
26. Wu, Q. *et al.*, 3D printing of microstructured and stretchable chitosan hydrogel for guided cell growth. *Adv. Biosyst.*, 1, 6, 1700058, 2017.
27. Akopova, T.A., Demina, T.S., Bagratashvili, V.N., Bardakova, K.N., Novikov, M.M., Selezneva, I.I., Istomin, A.V., *et al.*, Solid state synthesis of chitosan

and its unsaturated derivatives for laser microfabrication of 3D scaffolds. *IOP Conf. Ser.: Mater. Sci. Eng.*, 87, 1, 012079, 2015.
28. Almeida, C.R. *et al.*, Impact of 3-D printed PLA- and chitosan-based scaffolds on human monocyte/macrophage responses: Unraveling the effect of 3-D structures on inflammation. *Acta Biomater.*, 10, 2, 613–622, 2014.
29. Morris, V.B. *et al.*, Mechanical properties, cytocompatibility and manufacturability of chitosan: PEGDA hybrid-gel scaffolds by stereolithography. *Ann. Biomed. Eng.*, 45, 1, 286–296, 2017.
30. Demirtaş, T.T., Irmak, G., Gümüşderelioğlu, M., A bioprintable form of chitosan hydrogel for bone tissue engineering. *Biofabrication*, 9, 3, 035003, 2017.
31. Hong, N. *et al.*, 3D bioprinting and its *in vivo* applications. *J. Biomed. Mater. Res. Part B: Appl. Biomater.*, 106, 1, 444–459, 2018.
32. Zhao, Y. *et al.*, Three-dimensional printing of Hela cells for cervical tumor model in vitro. *Biofabrication*, 6, 3, 035001, 2014.
33. Norotte, C. *et al.*, Scaffold-free vascular tissue engineering using bioprinting. *Biomaterials*, 30, 30, 5910–5917, 2009.
34. Shi, W. *et al.*, Structurally and functionally optimized silk-fibroin–gelatin scaffold using 3D printing to repair cartilage injury *in vitro* and *in vivo*. *Adv. Mater.*, 29, 29, 1701089, 2017.
35. Xiong, S. *et al.*, A gelatin-sulfonated silk composite scaffold based on 3D printing technology enhances skin regeneration by stimulating epidermal growth and dermal neovascularization. *Sci. Rep.*, 7, 1, 1–12, 2017.
36. Das, S. *et al.*, Bioprintable, cell-laden silk fibroin–gelatin hydrogel supporting multilineage differentiation of stem cells for fabrication of three-dimensional tissue constructs. *Acta Biomater.*, 11, 233–246, 2015.
37. Michael, S. *et al.*, Tissue engineered skin substitutes created by laser-assisted bioprinting form skin-like structures in the dorsal skin fold chamber in mice. *PLoS One*, 8, 3, e57741, 2013.
38. Hospodiuk, M. *et al.*, The bioink: A comprehensive review on bioprintable materials. *Biotechnol. Adv.*, 35, 2, 217–239, 2017.
39. Guvendiren, M. *et al.*, Designing biomaterials for 3D printing. *ACS Biomater. Sci. Eng.*, 2, 10, 1679–1693, 2016.
40. Khalil, S. and Sun, W., Bioprinting endothelial cells with alginate for 3D tissue constructs. *J. Biomech. Eng.*, 131, 11, 111002, 2009.
41. Schuurman, W. *et al.*, Gelatin-methacrylamide hydrogels as potential biomaterials for fabrication of tissue-engineered cartilage constructs. *Macromol. Biosci.*, 13, 5, 551–561, 2013.
42. Rémy, M. *et al.*, Creation of highly defined mesenchymal stem cell patterns in three dimensions by laser-assisted bioprinting. *J. Nanotechnol. Eng. Med.*, 6, 2, 2015.
43. Li, J. *et al.*, 3D printing of hydrogels: Rational design strategies and emerging biomedical applications. *Mater. Sci. Eng.: R: Rep.*, 140, 100543, 2020.

44. Derakhshanfar, S. *et al.*, 3D bioprinting for biomedical devices and tissue engineering: A review of recent trends and advances. *Bioact. Mater.*, 3, 2, 144–156, 2018.
45. Compaan, A.M., Christensen, K., Huang, Y., Inkjet bioprinting of 3D silk fibroin cellular constructs using sacrificial alginate. *ACS Biomater. Sci. Eng.*, 3, 8, 1519–1526, 2017.
46. Vanaei, S. *et al.*, An overview on materials and techniques in 3D bioprinting toward biomedical application. *Eng. Regener.*, 2, 1–18, 2021.
47. Hu, F. *et al.*, Extrusion-based 3D printing of ceramic pastes: Mathematical modeling and in situ shaping retention approach. *Materials*, 14, 5, 1137, 2021.
48. Madduma-Bandarage, U.S. and Madihally, S.V., Synthetic hydrogels: Synthesis, novel trends, and applications. *J. Appl. Polym. Sci.*, 138, 19, 50376, 2021.
49. Dutta, S.D. *et al.*, 3D-printed bioactive and biodegradable hydrogel scaffolds of alginate/gelatin/cellulose nanocrystals for tissue engineering. *Int. J. Biol. Macromol.*, 167, 644–658, 2021.
50. Wei, J. *et al.*, Bioinspired 3D printable, self-healable, and stretchable hydrogels with multiple conductivities for skin-like wearable strain sensors. *ACS Appl. Mater. Interfaces*, 13, 2, 2952–2960, 2021.
51. Li, H., Tan, C., Li, L., Review of 3D printable hydrogels and constructs. *Mater. Des.*, 159, 20–38, 2018.
52. Zou, P. *et al.*, Temperature-sensitive biodegradable mixed star-shaped block copolymers hydrogels for an injection application. *Polymer*, 53, 6, 1245–1257, 2012.
53. Nie, L. *et al.*, Temperature-sensitive star-shaped block copolymers hydrogels for an injection application: Phase transition behavior and biocompatibility. *J. Mater. Sci.: Mater. Med.*, 24, 3, 689–700, 2013.
54. Fang, Y. *et al.*, Advances in 3D bioprinting. *Chin. J. Mech. Eng.: Addit. Manuf. Front.*, 100011, 2022.
55. Raveendran, S. *et al.*, Smart carriers and nanohealers: A nanomedical insight on natural polymers. *Materials*, 10, 8, 929, 2017.
56. Liu, F. *et al.*, Natural polymers for organ 3D bioprinting. *Polymers*, 10, 11, 1278, 2018.
57. Khan, M.J. *et al.*, Use of nanostructured polymer in the delivery of drugs for cancer therapy, in: *Nanostructured Polymer Composites for Biomedical Applications*, pp. 261–276, Elsevier, 2019.
58. Schmidt, L.M. *et al.*, Drug-loaded mesoporous silica on carboxymethyl cellulose hydrogel: Development of innovative 3D printed hydrophilic films. *Int. J. Pharm.*, 620, 121750, 2022.
59. Doppalapudi, S. *et al.*, Biodegradable polymers—An overview. *Polym. Adv. Technol.*, 25, 5, 427–435, 2014.

60. Lima, I.B. *et al.*, Development of nanostructured systems using natural polymers to optimize the treatment of inflammatory bowel diseases: A prospective study. *J. Drug Delivery Sci. Technol.*, 64, 102590, 2021.
61. Nie, L. *et al.*, Development of chitosan/gelatin hydrogels incorporation of biphasic calcium phosphate nanoparticles for bone tissue engineering. *J. Biomater. Sci. Polym. Ed.*, 30, 17, 1636–1657, 2019.
62. Niknam, F., Recent advances carbon nanomaterials as a new agent for breast cancer treatment. *Adv. Appl. NanoBio-Tech.*, 3, 1, 72–81, 2022.
63. Rajabi, M. *et al.*, Chitosan hydrogels in 3D printing for biomedical applications. *Carbohydr. Polym.*, 260, 117768, 2021.
64. Yadav, L.R. *et al.*, Chitosan-based 3D-printed scaffolds for bone tissue engineering. *Int. J. Biol. Macromol.*, 183, 1925–1938, 2021.
65. Taghizadeh, M. *et al.*, Chitosan-based inks for 3D printing and bioprinting. *Green Chem.*, 24, 1, 62–101, 2022.
66. Long, J. *et al.*, A 3D printed chitosan-pectin hydrogel wound dressing for lidocaine hydrochloride delivery. *Mater. Sci. Eng.: C*, 104, 109873, 2019.
67. Zhou, L. *et al.*, 3D printing of high-strength chitosan hydrogel scaffolds without any organic solvents. *Biomater. Sci.*, 8, 18, 5020–5028, 2020.
68. Sriphutkiat, Y. *et al.*, Cell alignment and accumulation using acoustic nozzle for bioprinting. *Sci. Rep.*, 9, 1, 1–12, 2019.
69. Abelseth, E. *et al.*, 3D printing of neural tissues derived from human induced pluripotent stem cells using a fibrin-based bioink. *ACS Biomater. Sci. Eng.*, 5, 1, 234–243, 2018.
70. Agostinacchio, F. *et al.*, *In situ* 3D printing: Opportunities with silk inks. *Trends Biotechnol.*, 39, 7, 719–730, 2021.

7

Electroactive Polymer Actuator-Based Refreshable Braille Displays

Pooja Mohapatra, Lipsa Shubhadarshinee and Aruna Kumar Barick*

Department of Chemistry, Veer Surendra Sai University of Technology, Siddhi Vihar, Burla, Sambalpur, Odisha, India

Abstract

On the report of the International Agency for the Prevention of Blindness's Vision Atlas, which compiles the most recent information on eye health, officially launched in 2021, there are 295 million people with middling-to-acute visual impairment and 43 million people who are unable to see worldwide. Those people require effective and affordable refreshable full-page Braille displays in order to take use of cutting edge digital technology and the numerous emerging computer-controlled devices. Blind or visually challenged people who only read in Braille typically employ a Braille monitor. In contrast, a refreshable Braille displays can be utilized as a computer add-on. Electroactive polymers (EAPs) are a particularly tempting replacement for low strain materials in many applications due to their huge strain, high elastic modulus, and ease of processing, which results in significantly better performances. In comparison to any other actuators created by humans, the EAPs are actuators that most closely resemble biological muscles, earning them the name artificial muscles. Owing to their ability, a variety of fascinating and innovative mechanisms, such as loudspeakers, robot fish, dust-wipers, and tiny grippers, can be produced and demonstrated. The test findings show that the EAP-based Braille actuator satisfies every functional standard employed for the actuators for refreshable full Braille displays, which provide portable dimension and affordable.

Keywords: Braille displays, Braille cells, Braille dot, electroactive polymer, actuators

**Corresponding author*: akbarick_chem@vssut.ac.in

Inamuddin, Tariq Altalhi and Ashjan Alrogi (eds.) Biomimicry Materials and Applications, (169–178)

7.1 Introduction

The statistics show that the blindness or vision impairment cause severe cruelty to millions of people. Public health policies, like resource earmark and service planning, are fundamentally based on current and precise data on the prevalence of blindness and visual impairment by aetiology. These data are also crucial for determining the importance of scientific breakthroughs and corporate research.

For people who are blind, Braille is a form of tactual reading and writing that they can use to communicate. It can be seen throughout history how the work of a 16-year-old boy, Louis Braille, in 1825, had a significant impact on reading and writing abilities of blind people. Captain Charles Barbier of France, who initially developed a 12-raised-dot code system to enable Napoleon's forces to communicate covertly at night, served as the inspiration for Braille. Braille condensed the system by employing each character having only six dots, placed as a rectangle pattern with two columns, each consisting of three dots. With such adjustment, deciphering the code became considerably quicker and much easier for fingerprints to perceive through touch. A subsequent variation includes two rows with eight dots each, bringing the number of feasible sign amalgamation from 64 to 256 i.e. 256 various Braille characters. Each dot's height relative to the reading surface must be individually adjustable. All readable American Standard Code for Information Interchange (ASCII) characters are represented by this code in a single cell [1]. Basically, the framework can be executed by a Braille cell containing eight standard dots as shown in Figure 7.1. On the implementation and removal of a voltage, the dot is downed and upped, respectively.

Approximately 338 million blind and visually impaired persons throughout the world are mostly left out of the information and communication technology revolution happening today. Indeed, text and

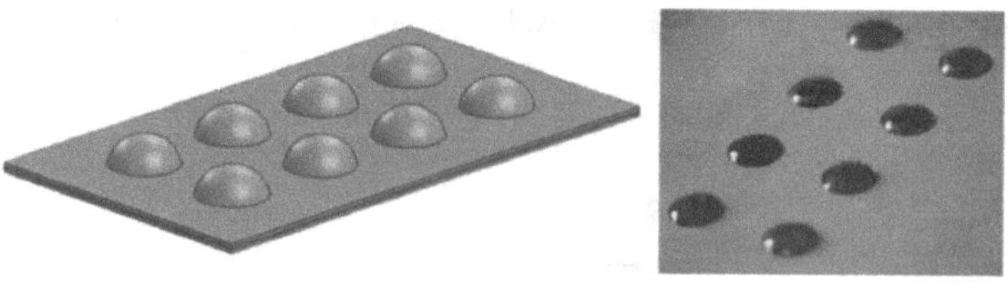

(a) (b)

Figure 7.1 Dynamic Braille cell representing (a) concept and (b) assemble prototype of array of eight Braille dots configuration based on dielectric elastomer actuators [2].

images can be delivered via visual sense owing to display on computers, mobile devices, touch screens, and other electronic gadgets. Only text-to-speech readers can provide access to digital informations for those with visual impairments. However, utilizing sound to communicate ideas is not always successful. In fact, the inability to continuously go back in time may restrict the comprehension of text based just on listening. Additionally, using headphones to maintain privacy or avoid disturbing other persons around may be necessary, and a busy location may present an extra difficulty. Therefore, the creation of refreshable Braille screens managed to somehow do away with these drawbacks. These are created as haptic interfaces that may be electronically programmed to allow blind users to go through Braille-coded content using lowered and dynamically raised dots. Full-page layouts in particular will be able to help blind individuals to ingress substantial amounts of organized and dynamic information using the sense of touch, just way people with healthy sight typically do with the perception of sight, for instance while operating smartphones, tablets, and computer monitors. We can claim that a refreshable Braille display, in contrast to a hardcopy version, must move a large number of dense dots quickly for simple understanding. These tactual displays could be devices presenting lone-character, single-line units, or character-filled full-page display boards. Based on piezoelectric reeds that activate the Braille dots, refreshable Braille displays are readily accessible in the marketplace. Figure 7.2 shows the commercially available refreshable Braille display.

Each cantilever on which the reeds are placed is equipped a Braille pin that rests on its open end. This technique restricts the total number of Braille characters on the display not more than two lines [4]. It prevents you from going back when reading a complete page of content. To get over this restriction, efforts are being made to create full-page Braille readers based on various kinds of piezoelectric actuators, but the only device that has been created so far is non-portable and is expected to cost around € 60,000 [2]. There are numerous kinds of materials that function according to various physical principles and exhibit noticeably diverse characteristics and abilities [5].

The size issue is resolved by limiting the active display to a single cell; however, the Braille cell cannot be traversed by the user's finger. Instead of reading as normal, the user rests their finger on the refreshable Braille cell. Prototypes and finished goods are typically instruments that hoist dots via holes on a leveled surface using electromechanical surface. Researchers have created and evaluated a variety of actuators and mechanisms over the years, including electromagnetic, shape memory alloy, which are the metals that can come back to their original formation before deformation such as piezoelectric, pneumatic, and thermal materials [6].

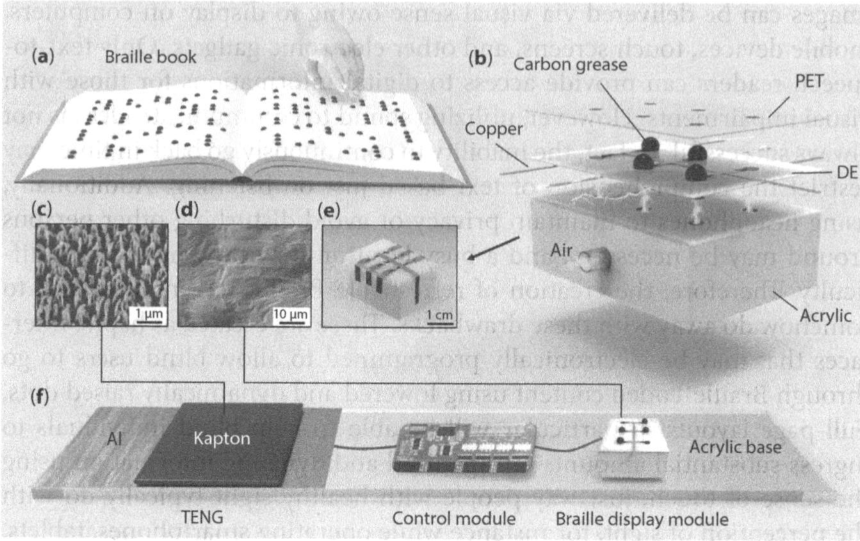

Figure 7.2 Photograph of the refreshable Braille displays system (RBDS): (a) Pictorial view of a standard Braille book, (b) Exploded View of the Braille Display Unit, (c) SEM Micrograph of the Surface of Kapton Film treated by Inductively Coupled Plasma (ICP) Reactive-ion Etching, (d) SEM micrograph of the carbon grease on the dielectric elastomer dots (DEDs), (e) picture of the Braille display unit, and (f) schematic illustration of the RBDS with control unit [3].

7.2 Refreshable Braille Display

The electroactive polymer (EAP)-based refreshable Braille display idea is based on the use of an array-shaped field-activated type EAP actuators. Individual elements in the array are activated by column on one side of an EAP film and rows of electrodes on the other. A Braille dot has been attached to each element, and by passing a voltage over the thickness of the chosen element, its local thickness reduces. Dots would be actuated by a computer to produce haptic patterns with highs and lows that represented the information to be read. As with reading traditional Braille print, a person would read by scanning with their fingertips. New Braille actuation technologies are definitely required to meet the need of former one. The Center for Braille Innovation (CBI), established by the National Braille Press, Boston, MA, USA, is trying to assist Braille technology innovators by assets and focusing attention in order to facilitate the creation of innovative electronic Braille displays. The CBI also gathers technical data about Braille, such as the following list of specifications for Braille display [4]:

Table 7.1 Details about Braille dot parameters recommended by European Blind Union [7].

Features	Dimensions (mm)
Height of dot	0.5
Diameter of dot	1.5
Vertical distance of Intra-cell	2.5
Horizontal distance of Intra-cell	2.5
Horizontal distance of Inter-cell	6.0

- Dot parameters (shown in Table 7.1)
- Dot height uniformity
- Dot displacement
- Dot force
- Braille cell thickness
- Dot setup time
- Acoustic noise
- Driving voltage and power
- Touch sensing
- Discrepancy between Braille and tactile graphics
- Reliability
- Maintenance
- Price
- The ability of electromechanically active polymers

7.3 Electroactive Polymers

An electroactive polymer (EAP) is a type of polymer that is affected by an electric field by altering its size or shape. Actuators and sensors are where this is most frequently used. Numerous EAPs have recently come into existence with the ability to support active, full-page displays [1, 8]. For the purpose of converting electrical energy into mechanical work for actuation, they are "smart" materials with an inherent capacity to change shape and/or size according to suitable electrical incentive. Additionally, they are capable of working in reverse, transforming mechanical energy into electrical form for mechanosensing i.e. the ability of the cell to feel mechanical indications of its microenvironment.

For researchers, the main advantage of using these materials is that they are able to jam more actuators into smaller spaces, avoiding interference with one other [8, 9].

The development of novel transducers based on EAPs is currently receiving more attention in order to fulfill the demand for Braille dot actuation technologies that can meet the criteria outlined above based on their shock resistance, mechanical compliance, lightweight, low mass density, easy processing, high scalability, acoustic noiseless, and low cost. EAPs are generally classified into two groups as mentioned below [4]:

- *(a)* ***Ionic EAPs:*** These are triggered by ion and/or solvent movement that is electrically stimulated and comprised of conjugated polymers, carbon nanotubes, polymer gels etc. giving cylinder or bender actuator configurations [10, 11].
- *(b)* ***Electronic EAPs:*** These are triggered by electrostatic forces and include electrostrictive polymers, piezoelectric polymers, liquid crystals elastomers and carbon nanotubes which show roll with spring core, bender with hydraulic and coupling, cylinder, diaphragm, hydrostatically coupled membranes etc configuration of actuators [12–17].

In consonance with various literatures, for energy density and actuation strain dielectric elastomers (DE) have high efficiency as compared to other EAPs [19]. Figure 7.3 shows the operational function of DE actuators.

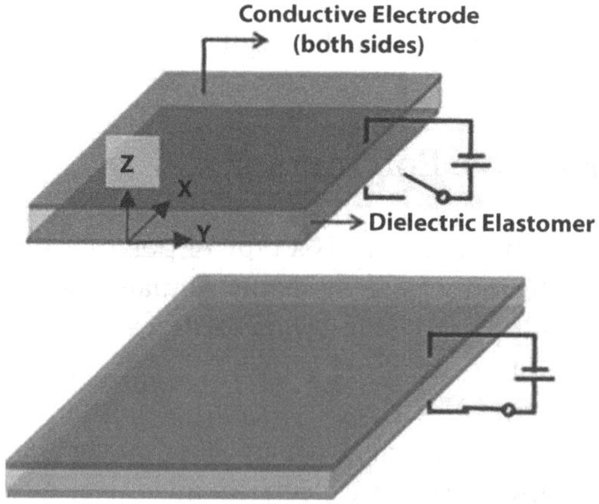

Figure 7.3 Working principle of a dielectric EAP before and after actuation, where the elastomer film is compressed in the thickness direction on the action of applied voltage across the electrodes [18].

Table 7.2 Psychophysical test result of adjacent Braille devices [15].

Subject number	Hit recognition rate (%)	Number recognition rate (%)
1	69	30
2	51	21
3	54	19
4	60	29
5	77	40
6	73	41
7	63	32

The field has reached a mature state as a result of the enormous advancement made in EAP science and technology over past 10 years, and serious efforts are now being made to design and build workable systems that would be suited for new refreshable Braille displays. The design of a water-impermeable surface should enhance the efficiency and long-term stability of the EAP. The possibility of losing positive outer ions will be limited by this design, when the operating EAP is engulfed in an aquatic circumstance and prevents the evaporation of water contained in the EAP.

The result of the psychophysical experiment with the Braille device can be seen in Table 7.2, where hit recognition rate (HRR) and number recognition rate (NRR) reach maximums of 80% and 41%, respectively. These two are the rates at which two adjacent Braille cells are read by the data and the subject. HRR represents the rate at which a subject acknowledges Braille dot movements, and NRR is used to regulate whether a subject recognizes the character in the right way or not [15].

7.4 EAP-Based Braille Actuator

An experimental example of Braille actuator was created by winding terpolymer sheets around a prestressed narrow spring with thin conductive polymer electrodes. This fabrication method is represented in Figure 7.4.

For refreshable Braille displays, Chakraborty *et al.* [18] chose a minimalist dielectric elastomer tubular actuator (DETA). Accordingly, EAP appears to offer the most potential for achieving the current requirements

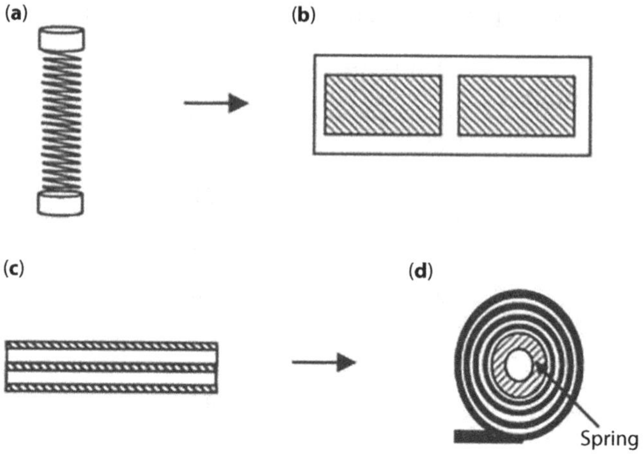

Figure 7.4 Illustrative representation of EAP actuator for Braille display during the fabrication process using the electrostrictive polymer thin films those can be mentioned orderly as: (a) pre-strained spring core, (b) electrode with polymer thin film, (c) to separate the positive and negative ends, thin film is folded to form a bilayer, and (d) actuator that is rolled [12].

for refreshable Braille displays, despite the fact that a broad range of materials have the capacity of creating actuation in response to an extrinsic influence. The concept incorporates a straightforward, very simple-to-manufacture DETA under the best prestrain. In an electric field, to raise the Braille dot to the necessary elevation, the improved prototype displayed sufficient shift. Figure 7.5 displays the EAP Braille actuator optic image [9].

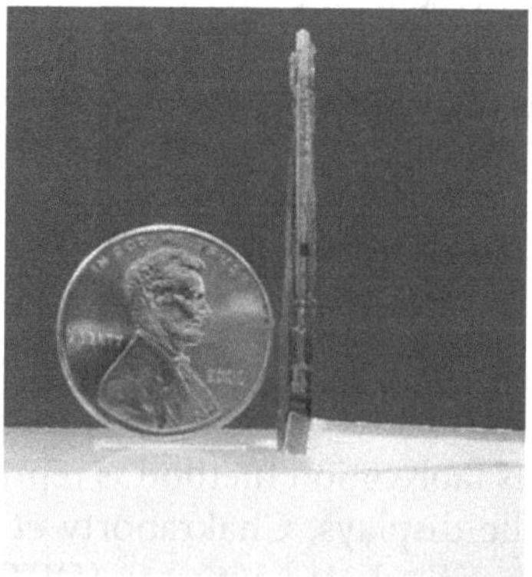

Figure 7.5 EAP Braille actuator optic image [12].

7.5 Conclusions

In recent decades, it has been recommended to implement a revolutionary novel Braille cell design for a full-page refreshable Braille display. The obstacle in creating such tactile interfaces is finding the right actuation technology that can drive each and every dot in every single cell of the entire display while using as little mass, size, energy, and money possible. In order to achieve this a wide range of actuation technologies, including electromagnets, shape memory alloys. Bimetals, piezo ceramic reeds, pneumatic drives, and many others, have been thoroughly scrutinized. While real-size refreshable Braille dots based on electroactive polymers is an attempt to overcome the shortcomings of all existing cutting edge techniques. The creation of a full-fledged refreshable Braille display is crucial for the inclusion of the blind in this modern communication epoch.

References

1. Bar-Cohen, Y., Electroactive polymers for refreshable Braille displays. *SPIE Newsroom*, *11*, 1–3, 2009.
2. Frediani, G., Busfield, J., Carpi, F., Enabling portable multiple-line refreshable Braille displays with electroactive elastomers. *Med. Eng. Phys.*, *60*, 86–93, 2018.
3. Qu, X., Ma, X., Shi, B., Li, H., Zheng, L., Wang, C., Liu, Z., Fan, Y., Chen, X., Li, Z., Wang, Z.L., Refreshable braille display system based on triboelectric nanogenerator and dielectric elastomer. *Adv. Funct. Mater.*, *31*, 5, 2006612, 2021.
4. Runyan, N.H. and Carpi, F., Seeking the 'holy Braille' display: Might electromechanically active polymers be the solution? *Expert Rev. Med. Devices*, *8*, 5, 529–532, 2011.
5. Brochu, P. and Pei, Q., Dielectric elastomers for actuators and artificial muscles, in: *Electroactivity in Polymeric Materials*, L. Rasmussen (Ed.), Springer, Boston, MA, 2012.
6. Runyan, N. and Nassimbene, E., Alphanumeric 'displays' for the blind: A technology search. *IBM Technical Report*, IBM Corp., Los Gatos, CA, pp. 1–33, 1974.
7. Leonardis, D., Claudio, L., Frisoli, A., A survey on innovative refreshable braille display technologies, in: *Advances in Design for Inclusion*. AHFE 2017. Advances in Intelligent Systems and Computing, G. Di Bucchianico, and P. Kercher (Eds.), vol. 587, Springer, Cham, 2018.

8. Bar-Cohen, Y., Electroactive polymer actuators as artificial muscles, in: *Biomimetics: Nature Based Innovation*, Y. Bar-Cohen (Ed.), Taylor and Francis, Boca Raton, FL, 2012.
9. Bar-Cohen, Y., Active reading display for blind (ARDIB). *Submitted as a New Technology*, 2001.
10. Spinks, G.M. and Wallace, G.G., Actuated pins for braille displays, in: *Biomedical Applications of Electroactive Polymer Actuators*, F. Carpi, and E. Smela (Eds.), pp. 265–277, Wiley, NY, 2009.
11. Kato, Y., Sekitani, T., Takamiya, M., Doi, M., Asaka, K., Sakurai, T., Someya, T., Sheet-type Braille displays by integrating organic field-effect transistors and polymeric actuators. *IEEE Trans. Electron. Devices, 54*, 2, 202–209, 2007.
12. Ren, K., Liu, S., Lin, M., Wang, Y., Zhang, Q.M., A compact electroactive polymer actuator suitable for refreshable Braille display. *Sens. Actuators A: Phys., 143*, 2, 335–342, 2008.
13. Di Spigna, N., Chakraborti, P., Winick, D., Yang, P., Ghosh, T., Franzon, P., The integration of novel EAP-based Braille cells for use in a refreshable tactile display, in: *Proceedings of SPIE 7642, Electroactive Polymer Actuators and Devices (EAPAD)*, pp. 53–61, 2010.
14. Heydt, R., Refreshable Braille display based on electroactive polymers, in: *Proceedings of 23rd International Display Research Conference*, Phoenix, Arizona, 2003.
15. Choi, H.R., Koo, I.M., Jung, K., Roh, S.G., Koo, J.C., Nam, J.D., Lee, Y.K., A Braille display system for the visually disabled using a polymer based soft actuator, in: *Biomedical Applications of Electroactive Polymer Actuators*, F. Carpi, and E. Smela (Eds.), pp. 427–442, Wiley, NY, 2009.
16. Yu, Z., Yuan, W., Brochu, P., Chen, B., Liu, Z., Pei, Q., Large-strain, rigid-to-rigid deformation of bistable electroactive polymers. *Appl. Phys. Lett., 95*, 19, 192904, 2009.
17. Carpi, F., Frediani, G., De Rossi, D., Opportunities of hydrostatically coupled dielectric elastomer actuators for haptic interfaces, in: *Proceedings of SPIE 7976, Electroactive Polymer Actuators and Devices (EAPAD)*, pp. 364–372, 2011.
18. Chakraborti, P., Toprakci, H.K., Yang, P., Di Spigna, N., Franzon, P., Ghosh, T., A compact dielectric elastomer tubular actuator for refreshable Braille displays. *Sens. Actuators A: Phys., 179*, 151–157, 2012.
19. Pelrine, R., Kornbluh, R., Pei, Q., Joseph, J., High-speed electrically actuated elastomers with strain greater than 100%. *Science, 287*, 5454, 836–839, 2000.

8

Materials Biomimicked From Natural Ones

Carlo Santulli

Geology Division, School of Science and Technology, Università di Camerino, Camerino, Italy

Abstract

Biomimicking materials from natural ones poses a number of problems, although it is the most direct approach to bioinspiration, compared to the bare extraction of design concepts from biological organisms and applying them to another design without any consideration of geometrical hence structural similarity. The issues concentrate particularly on the achievement of an adapted material geometry, able to effectively reproduce the spiraliform effect and the subsequent layering philosophy, experimented in nature. In practice, only a few species have been specifically targeted for the purpose, following mechanical studies and observation campaigns. This chapter concentrates specifically on giving a brief account of the principal achievements obtained as regards mechanical performance characteristics not currently available in engineering design of materials. Examples of these are the coexistence of high toughness, hardness, and strength in ceramics, the possibility to modify performance under stress of the same material only by structure design and the obtainment of reversible adhesion regardless of substrate.

Keywords: Nacre, spider silk, gecko, damage tolerance, self-cleaning, adhesion

8.1 Introduction

A material can be defined as any 3-D assembly that can withstand a mechanical load, even if it was its own weight. Nature has developed its own materials, to build up the specific structures that make their organisms (animal, vegetable, or mineral) possible. Natural materials are arranged according to a hierarchical structure, so it is self-assembling during design and realized

Email: carlo.santulli@unicam.it

Inamuddin, Tariq Altalhi and Ashjan Alrogi (eds.) Biomimicry Materials and Applications, (179–198)

into spiral forms to allow for their development. This characteristic is due to the fact that natural materials typically include a limited number of components with intrinsically scarce properties; however, the complexity of their architectures allows compensating for that when growing over several order of magnitudes [1]. To grow engineered materials from biological processes would involve their use to obtain some specific characteristics, with no need to synthetically develop materials that are able to match as closely as possible the natural properties, but without copying the natural strategy of production. So far, this possibility appears a rare occurrence for a number of limitations, in particular, the fact that biological structures are supposed to work in a restricted range of temperatures, and synthesis starts from the single cell to allow maintaining the possible characteristics of repair, growth, or replenishment [2].

In other words, the construction of materials in nature has no regard whatsoever for the possible obtainment of linear forms, which would be static and impede any development or repair: this principle extends also to ceramic materials, such as calcium carbonate used for the construction of coccoliths [3], or silica nanostructures used for diatoms [4]. While the spiral arrangement is functional for the needs of natural organisms, it significantly complicates the reproduction of natural materials by traditional engineering techniques, also because their properties are strictly linked to their surface arrangement. The availability of additive manufacturing techniques most recently allow the form of the material to be loosened from most stringent technological constraints, and move towards a customized design, a method which identifies the uniqueness of natural structures [5]. This applies both to the material's surface and to its inner structure, where the balance of full and empty spaces and the shapes of the cells in the lattice play an essential role [6], as indicated in Figure 8.1.

Reproducing this mode of design requires other than microscopical observation (scanning electron microscopy, atomic force microscopy) and nanometric testing, such as nanoindentation [7], also X-ray based techniques, such as microcomputed tomography [8]. This is of substantial interest in many engineering applications because it allows obtaining other properties, which are important in terms of functionality, such as the possibility of repair without the addition of further material. Once the microscopic structure of the natural material is observed and recognized, different possibilities open. An option is to directly copy it, which has been referred to as "biotemplating." This is a method that concentrates on the precise reproduction, e.g., of the hierarchical arrangement present in a natural material then transferring it to a synthetic one. One of the major successes of this method has been in the reproduction of the *Morpho*

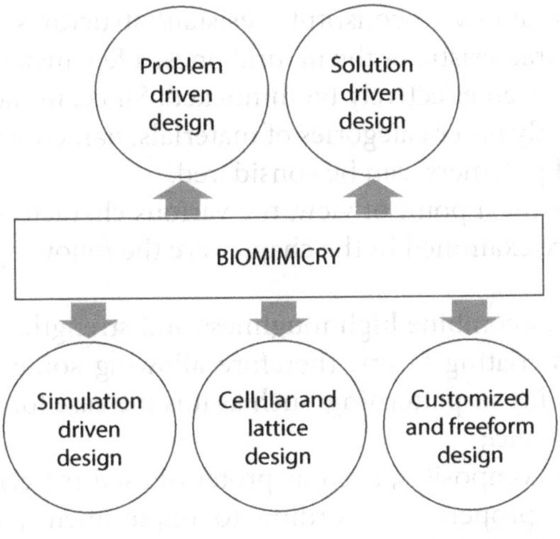

Figure 8.1 Design and solutions for biomimicry of materials [6].

rhetenor butterfly wings' patterns, in order to reproduce their structural color capabilities for application in photocatalysts [9]. As regard mechanical performance, this has been performed for example on *Luffa cylindrica* porosity features, then reproducing them into zeolite materials, with the idea of obtaining catalytic reactors [10]. In other way, it can be possible to simply extract more generally the inherent design principles encoded and then fabricating novel bioinspired materials [11].

Therefore, a reason to take inspiration from natural materials in order to develop engineered substitutes while maintaining some specific properties, which is the objective of biomimicking them, has been triggered by the desire to reduce the energy used for the purpose. This implies rather relying on microscopic surface and internal structure modifications, and not on the macroscopic features of the components to confer to the structure the desired characteristics.

In general terms, advantages in using bioinspiration for the development of materials can be classified in different categories, as from a suggestion exposed in [12]. The expected gains can be from an electrical, biological, chemical, or mechanical point of view, or multifold, and possibly involve an advantage in terms of sustainability.

This chapter would consider a few cases of biomimicking natural materials, while making considerations on the potential these applications offer, and on the limitations they still present; in other words, this would concentrate on those purposes and aims for biomimicking that uniquely pertain to the mechanical domain. In practice, concentrating on those characteristics that resort in an ability to constitute resistant structures with properties that are only characteristic of the natural ones, a few materials can be recognized to have been effectively biomimicked. Since, in nature, no metals are developed, only other categories of materials, namely ceramics, fibrous composites, and polymers, can be considered.

From a mechanical point of view, the various characteristics of natural materials that are examined in the chapter are the following:

- Ceramics combine high toughness and strength, also when used as coating layers, therefore allowing some degree of workability or patterning, such as it is the case for *nacre* and *tooth enamel*;
- Fibrous composites, such as protein-based materials, allow variable properties according to requirements, as in the case of *dragline silk*, or of *fish scales*, which dissipate energy through a plywood-like mechanism;
- Polymers are fit for easy, durable, and reversible junction, which includes self-cleaning abilities in difficult situations (i.e., under water, or when undergoing oil/grease contamination) with other materials, such as in the case of *gecko feet*, *blue mussel*, and *sharkskin*.

In the following sections, a few examples of successful biomimicking from natural materials are reported more into detail, giving some indications about the reproducibility of the phenomenon and its prospected applications.

8.2 Damage-Tolerant Ceramics

8.2.1 General Considerations

Manufacturing ceramics with a combination of properties including high toughness, high hardness and high strength does represent an issue, since these properties might typically conflict among themselves, though their coexistence guarantees to the material a real and effective "damage tolerance." Superhard materials would be desirable for some applications, such as for cutting tools, while on the other side they need to be sufficiently tough as to avoid brittle failure for very fast crack propagation: in this sense, it has been suggested that functional grading, together with the availability of additive manufacturing, would make the production of materials with both characteristics feasible [13]. Nature realizes this without particular problems: such characteristic of natural materials has been attributed to their hierarchical structure, which is able to ensure that during loading the ceramic at the microscale, hence "intrinsically," departs in a significant way from a linear elastic behavior [14]. This is typical of natural ceramic composites, such as in sponge spicules, constituted by hydrated silica bound together by protein (silicatein) [15]. The structural properties of sponge spicules have suggested earlier applications as biomimetic optical fibers, able especially to be used under flexural loads for their aforementioned composite arrangement [16].

8.2.2 Nacre

A typical example of damage-tolerant ceramic material is nacre. A consideration that appears common knowledge in recent materials engineering is that to achieve an effective damage tolerance the construction of a layered structure would be successful also in the case of ceramics, where in practice it would deflect the crack in its propagation across the material through the combined use of a cross-lamellar structure [17]. This makes the ceramic composite not only able to survive large impact phenomena, which is the typical sense of fabrication of composite structure to promote dissipation of energy through the occurrence of damage, but also offers to it the capability to withstand localized indentation load. The "armor" metaphor for some biological structures, such as shrimp shells, therefore their design potential for ballistic impact resistance, has been explored for possible engineering reproduction already in [18]. This has also been indicated in conch (*Strombus gigas*) calcium carbonate shells, where the hierarchical structure is able to provide a prevalent orientation at ±45°, with actual similarity to a quasi-isotropic polymer matrix composite, yet with a much harder and tougher intrinsic structure [19].

This has been proved, e.g., by a biomimicking approach in complex ytterbium-aluminum oxides, particularly for use as heat barriers in view of their low thermal conductivity and thermal expansion close to nickel superalloys [20]. However, outstanding surface characteristics, such as hardness, are not achievable other than by the progressive and spiraliform deposition of a very large number of layers. In this way, the thickness of the single layer would be inferior by several orders of magnitude to that of the whole structure, such as in seashells, therefore replicating the brick-mortar effect on the nanoscale [21]. The first studies on the shell of the red abalone (*Haliotis rufescens*) indicated, contrary to the largest majority of ceramic materials, a large dependance of strength on its orientation, and a significant sensitivity to the strain rate of the behavior obtained following load application [22]. As a matter of fact, the structure of nacre is especially studied with respect to the construction of the whole shell and is constituted by microtiles of the hard component, aragonite, arranged together by a soft polysaccharide (chitin) and protein-based component [23]. The material is arranged as a combination of solid nacre with growth layers, as reported in Figure 8.2, which have a particular shear and tensile behavior in the hydrated state allowing for larger deformations and more flexibility [24].

Nacre-like components are arranged into hexagonal platelets, which can be mimicked using appropriate Voronoi cells characteristics, in terms of radiuses and angles, so that seeds are randomly shifted. This was reproduced using polylactic acid as the stiff material for platelets, coupled then with softer materials, such as thermoplastic polyurethane (TPU) or nylon [25]. This has also been biomimicked by the immersion of a glass substrate into two polyelectrolyte solutions, namely poly(acrylic acid) (PAA) and poly (4-vinyl pyridine) (PVP), then dissolving PAA at pH=10 and forming nanometric pores, followed by quenching and stabilization by UV crosslinking and final immersion again into PAA solution [26]. Another possible replica was obtained based upon graphene (G) coupled with molybdenum disulfide (MoS_2) and thermoplastic polyurethane (TPU) in a ternary nanocomposite. In this case, self-assembly would be based on a graphene nanosheet/organic matrix manufactured as an alternating layered structure, thus establishing interactions at the interface that serves as the boundary between graphene nanosheets and organic components [27].

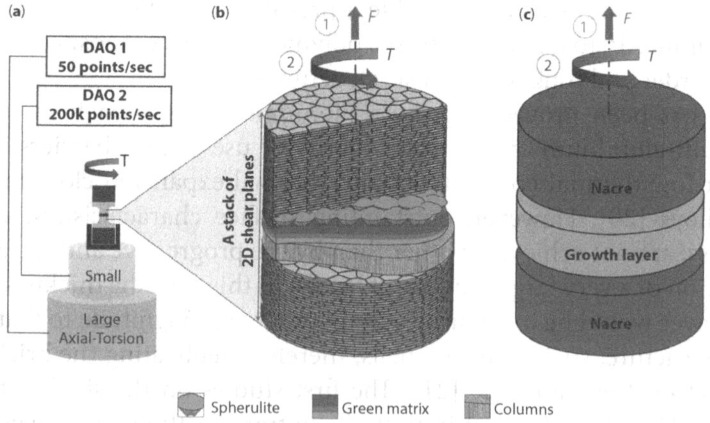

Figure 8.2 Structure of red abalone and model of it under tensile and shear loading [24].

8.2.3 Tooth Enamel

The process that leads to the very high hardness and strength to the application of different forces, especially shear and compression, of tooth enamel, is based on the mineralization of a proteinaceous matrix, which is able to be achieved by judicious modifications in packing and composition, to get to a calcium phosphate (hydroxyapatite) content of up to 95 wt. %, as reported in [28]. This is the highest ceramic content obtained in a natural structure, with absence of collagen and appears particularly difficult to be replicated using a practically achievable and reproducible route [29]. The biomineralization process of tooth enamel has been tried to be successfully replicated in vitro [30]. Amelogenesis i.e., the formation of enamel of teeth, comprises three phases: protein assembly, proteolytic digestion, and crystal growth. The reproduction of this process has a particular interest for dentistry, which has been based so far on substitutive materials. Five possible pathways appear possible for the purpose, namely physicochemical synthesis, protein matrix-guided crystal growth, which is led by the action of protein matrix, the remineralization of surface, cell-based engineering of the structure starting at the cell level, and finally the biological regeneration [31]. Remineralization in particular has been provided by applying hydroxyapatite (HA) dendrimers, i.e., polymeric trees, to provide the formation of the nanorod-like HA arranged into highly organized prisms, as required by tooth enamel biosynthesis [32].

8.3 Protein-Based Materials With Tailored Properties

8.3.1 General Considerations

Obtaining different properties, therefore usability for different purposes, with a material that has potentially always the same properties and even similar dimensional characteristics, has the important consequence to allow biomimicry entering fields, such as the production of textiles. Potential applications can develop considering the number of assets offered by natural organisms, which can be of use in this field, among which are functional surfaces, structural colors, self-healing potential, and thermal insulation capabilities [33]. This is because the increase of properties in these materials, maintaining the same base material, has always been achieved by modifying the weight per square meter, by changing the weave architecture, passing from plain to twill or satin, etc., or modifying, in most cases by chemical treatments, the surface properties of the fibers. Of course, these methods can also be combined. However, nature does not use this perspective e.g., since the change of dimensions is not always a possible option, since size mainly depends on the environmental context in which the structure develops, and chemical modification post-extrusion would be cumbersome if not impossible at all.

Initial studies on application of biomimicry to textiles mentioned as possible wire-developing species for inspiration are spiders, hagfish slimes and mussels [34]. Despite the fact that practical application of spiders as direct silk producers for biomimetic purposes is difficult to think of because, among other things, of their proneness to cannibalism, it is important to highlight that spider silk appears most interesting from the mechanical point of view. The fibers are normally extruded from a nozzle of fixed size, which is placed in the mouth in the case of spider silk, while this comes from the foot in the case of mussel byssus. To be able to modify the behavior of the textile without changing its dimensions, different approaches can be followed: in the case of spiders producing dragline silk, the extrusion velocity is changed according to requirements [35]. However, also other parameters are involved, such as dope composition, hence, the protein feedstock for spinning, temperature, pH, and metallic ions [36]. In contrast, for mussel byssus, fiber properties are different according to their position in the structure, more specifically extensibility is offered by the coiled fibers with loose packing located in the proximal region, while the distal bundles of filaments with their higher density provide the necessary stiffness [37].

8.3.2 Dragline Silk

In the case of dragline silk, the main interest is correlated with the fact that from a single liquid protein (spidroin), it is possible to obtain materials with a wide range of properties. In particular, it is possible to range from comparatively stiff and strong dragline silk, therefore with moderate strain at failure, to seizure silks with reduced strength, yet able to achieve very high toughness and huge strains to failure [38]. A first approach concerned the production of artificial spider silk using recombinant or reconstituted spidroins and spinning techniques. Recombination starts from the observation of amino acid sequences of full length spidroin: this is tripartite, including at the extremes two non-repetitive domains with amino (N) terminal and carboxyl (C) terminal, respectively, and a central highly repetitive central part including around 100 polyalanine/glycine rich co-segments [39]. It is suggested that the two terminals need to be included in any case, as they have different functions. For this reason, spidroin can be miniaturized by linking the terminal domains with a limited number of repetitive segments, which can reproduce sufficiently the properties of native spider silk.

This will only work effectively in the case the tripartite structure synthesized is able to maintain water solubility until the activation of fiber formation from the liquid protein occurs. In particular, self-assembly of recombinant spidroins allows obtaining a toughness not far from native spider silk [40].

8.3.3 Fish Scales

Fish scales present a structure that allow particular ductility in withstanding different types of loads, such as indentation, tensile and tear: this is enabled by their two components, the protein-based extracellular matrix, formed by type I collagen fibers, and their ceramic reinforcement formed by hydroxyapatite (HAP) with low amount of calcium [41] (Figure 8.3). In dimensional terms, specific studies have been for instance carried out on the *Arapaima gigas* (piracucu) scales, where the internal lamellae show a thickness in the region of 50 to 60 μm, whereas the collagen fibers are much finer, having around 1 μm diameter each [42]. They also prove particularly suitable to develop scaffolds for tissue engineering applications in particular species [43]. This is the case, e.g., for grass carp, where multiple collagen sheets are arranged in a structure, resembling ultimately to cornea, cementum, and tibial matrix, with a double-twisted spiral architecture reproducing a liquid crystal [44]. Fish scales have also been studied from the biomimicry point of view with the idea to obtain super-oleophobic properties, in particular reproduced from the filefish scale surface [45].

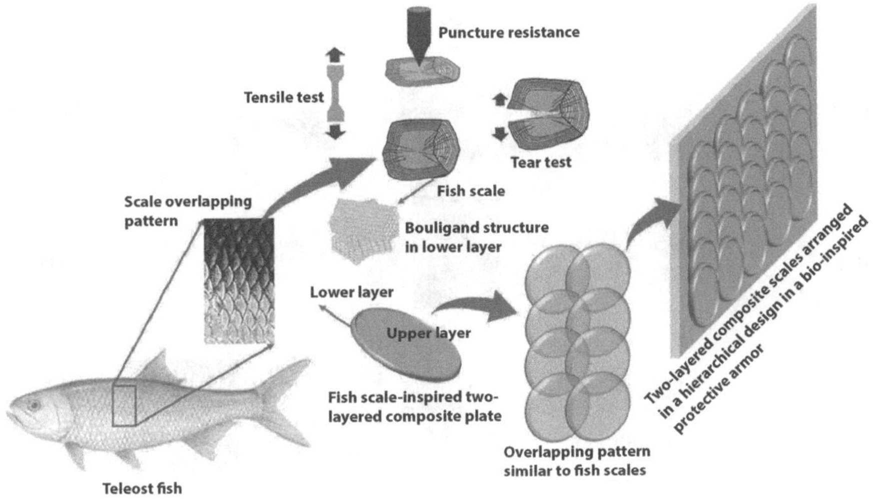

Figure 8.3 Structure of fish scales to form two layered composites with overlapping patterns [41].

8.4 Polymers Fit for Easy Junction/Self-Cleaning

8.4.1 General Considerations

As recent studies have pointed out, the ability of a natural structure to join with other materials flexibly and reversibly is correlated with its self-cleaning potential, which is usually associated with lotus (*Nelumbo nucifera*) leaf characteristics. This correlation does exist because it effectively responds to the same requirement of not being contaminated or else adapt to the actual situation of contamination existing on the surface with different and tailored characteristics. This is reported in Figure 8.4, where various species that show the different characteristics are described together

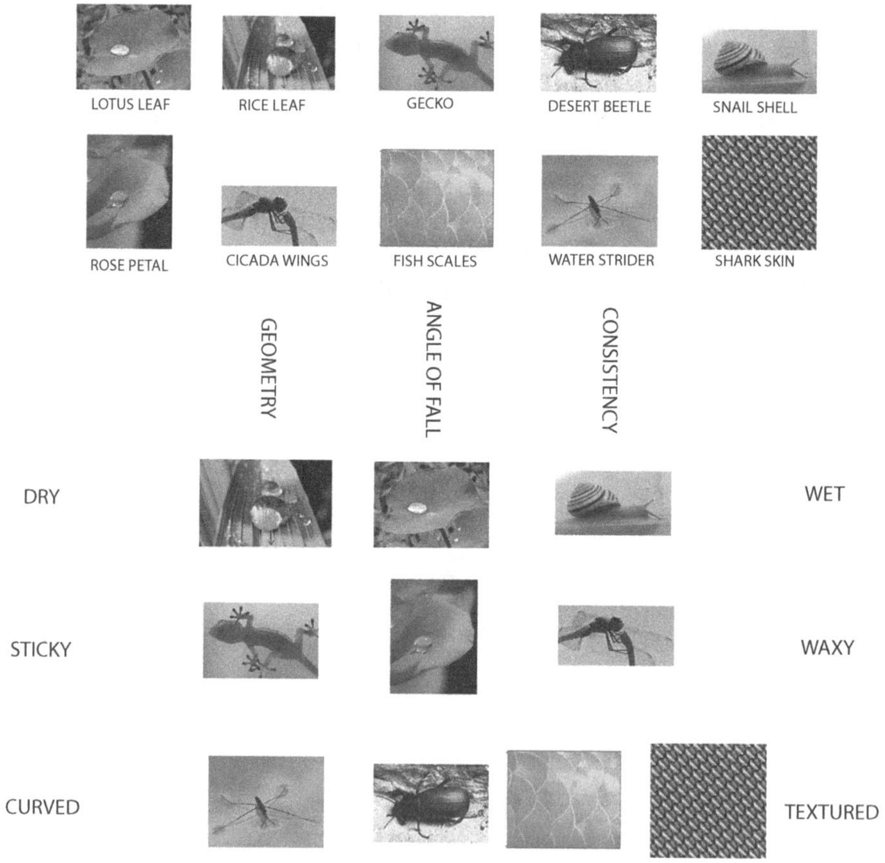

Figure 8.4 Different species with self-cleaning/textured adhesion capabilities (original property).

with the properties of the surfaces/environments with which they have to interact [45, 46]. To biomimic superhydrophobic surfaces, combining the properties from lotus leaf and silver ragwort leaf (*Jacobaea maritima*), electrospun polystyrene fibers from *N'N*-dimethylformamide (DMF) and different amounts of nanosilica particles have been used [47]. More in general, the selected approach is to mimic superhydrophobic surfaces through arranging hydrophobic materials into rough substrates [48]. In general terms, to biomimic well-known superhydrophobic structures, such as lotus leaf and rose petal, it is essential that polymer topography for the reproduction of the effect is carefully tailored through chemical modification, e.g., by allowing the reaction of branched polymers with appropriate alkali metal ions [49].

8.4.2 Gecko for No-Glue Adhesion

Geckoes toes are able to provide a "glueless" adhesion effect, which is adaptive to the different types of habitats in which they live, therefore, a number of possible substrates for safe and effective, but especially reversible, gripping [50].

Poly(glycerol-co-sebacate acrylate) (PGSA) has been purposely modified on its surface texture, so to provide an adhesive not only biodegradable and biocompatible, yet also complying with mimicking gecko feet's

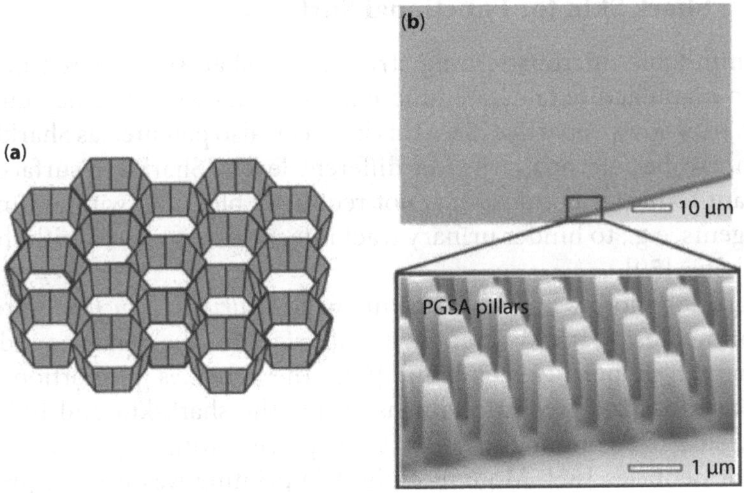

Figure 8.5 (a) Accordion like structure and (b) SEM image of incorporated gecko inspired pillar patterns on PGS matrix [53].

nanotopography [51]. An effective reproduction of it has been obtained by an "accordion-like" structure, declined into pillars, as reported in Figure 8.5 [52]. The pillar structure shows limitations nonetheless in trying to reproduce the self-cleaning characteristics, namely in dry environments, offered by the presence of a hierarchical fibrillar structure in gecko's feet [53].

8.4.3 Blue Mussel for Development of Specific Adhesives

The adhesive proteins of the blue mussel are developed by a high proportion of some amino acids, such as 3,4-dihydroxyphenylalanine (DOPA), together with minor quantities of glycine and L-lysine, modified enzymatically following tyrosine biosynthesis [54]. In particular, the centrality of the oxidase-mediated polymerization of quinones in the adhesion strategies of blue mussel is suggested [55]. These adhesive peptides were also synthesized by chemo-enzymatic polymerizations, using papain and tyrosinase, then enhancing the effectiveness of synthesizing peptide-based materials, such as Mefp-5 (*Mitilus edulis* foot protein n.5) [56]. In particular, the same Mefp-5 was synthesized and proved effective for adhesion with complex substrates, intended for sub-water use, such as mica, and also verified against different salt contents: the results indicated that Mefp-5 (30% mol. of DOPA) has an adhesion strength about 5 times superior against another adhesive protein, Mefp-3 [57].

8.4.4 Shark Skin for Functional Surfaces

The inimitable micropatterning structure of sharkskin offered inspiration to manufacture fabricate functional surfaces with specific fluid (air, water, etc.) flow properties [58]. This has been also patented as Sharklet™, which can be micropatterned at different levels. Sharklet's surfaces are resistant to bacterial infections, not requiring blending with antimicrobial agents, e.g., to hinder urinary tract infections associated with the use of catheter [59].

Sharkskin models from the shortfin mako (*Isurus oxyrinchus*) were also 3-D printed using two different materials, rigid for the denticles and flexible for the supporting membrane [60]. The denticles' proportional size was scaled up to a realistic dimension for the sharkskin and indicated an increase of swimming speed by a few percent with respect to a model without denticles [61]. More recently, 3-D printing was used to produce superoleophobic shark skin to offer a sustainable supply of energy by mechanical–electrical conversion through the mechanically sliding energy from droplets [62], as reported in Figure 8.6.

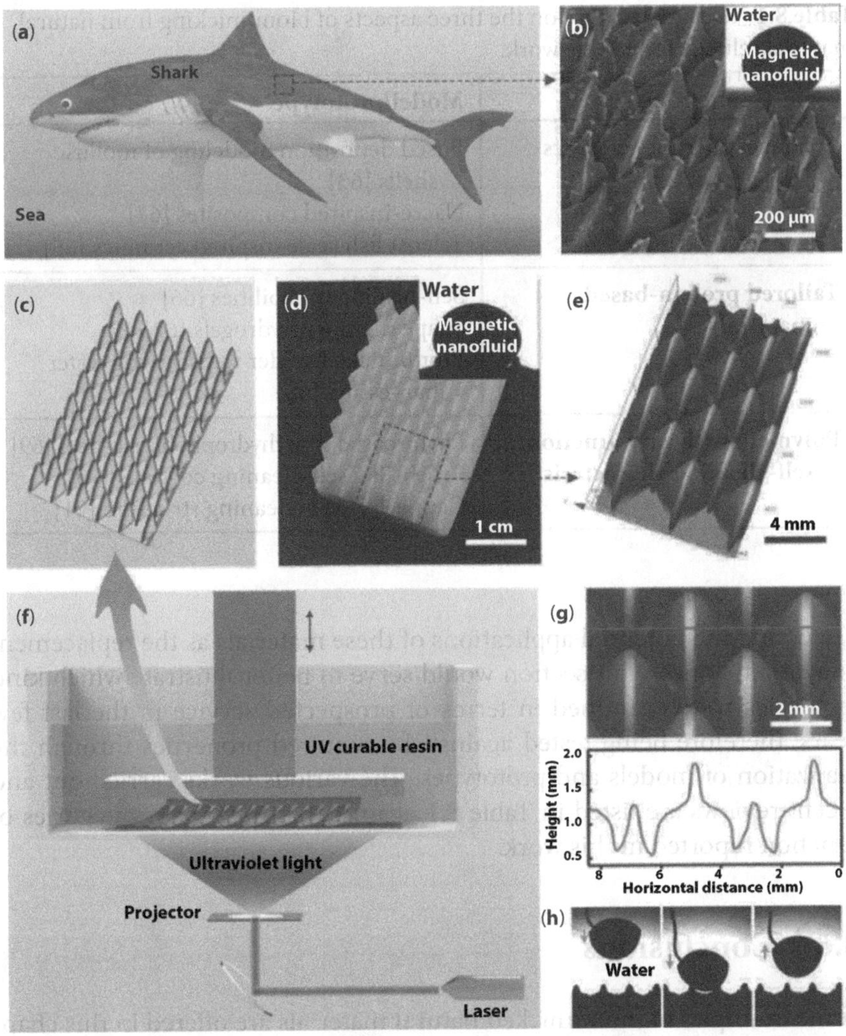

Figure 8.6 3D-printed sharkskin offering superoleophobic and low adhesion properties underwater: (a) Shark scheme. (b) Scanning electron microscopy (SEM) image of the sharkskin with relevant static contact angle for magnetic nanofluid droplet. (c), (d) and (e) Phases of design (f) Additive manufacturing system (g) Magnified image of completed design in (e), with height curve at the section line. (h) Adhesive force underwater [61].

8.5 Recent Prototype Developments on Materials Biomimicked from Natural Ones

After having illustrated the different characteristics of materials biomimicked from natural ones, the results can be summarized by briefly

Table 8.1 Recent literature on the three aspects of biomimicking from natural materials elucidated in this work.

Type of function	Model/prototype
Damage-tolerant ceramics	Fused deposition modeling of mollusc shells [63] Nacre-inspired composites [64] Teleost fish scale inspired ceramics [65]
Tailored protein-based materials	Self-healing capabilities [66] Shape memory hydrogels [67] Biomimicked spider webs for fog water harvesting [68]
Polymers with easy junction/ self-cleaning characteristics	Phyto-based superhydrophobic surfaces [69] Brush-like self-cleaning coatings [70] Canna-leaf self-cleaning structure [71]

discussing the potential applications of these materials as the replacement of synthetic ones. This section would serve to better illustrate which kind of species were examined in terms of prospected service in the last few years, therefore being tested against the required properties through the realization of models and prototypes. The various works carried out and recent reviews are listed in Table 8.1 against the three main categories of function reported in this work.

8.6 Conclusions

Some examples of biomimicked natural materials are offered in this chapter: this does not pretend to be exhaustive, yet to offer some indications on the trends followed on using bioinspiration from nature on materials, which revolve around a few themes, in particular damage tolerance, properties tailoring, self-cleaning and easy junction. In this regard, some species have been definitely paradigmatic, such as it is the case e.g., for red abalone, spiders, lotus leaves, geckoes, blue mussels, sharks. It can be considered though that the refinement of the problem-solving approach applied in biomimicry, coupled with a multidisciplinary approach to the species, able to encompass biological knowledge with engineering requirements, would give space to further bioinspiration considerations on other families of animals and plants.

References

1. Wegst, U.G., Bai, H., Saiz, E., Tomsia, A.P., Ritchie, R.O., Bioinspired structural materials. *Nat. Mater.*, 14, 23, 2015.
2. Reed, E.J., Klumb, L., Koobatian, M., Viney, C., Biomimicry as a route to new materials: What kinds of lessons are useful? *Philos. Trans. R. Soc. Lond. A*, 367, 1571, 2009.
3. Mann, S., The chemistry of form. *Angew. Chem. Int. Ed.*, 39, 3392, 2000.
4. Yang, W., Lopez, P.J., Rosengarten, G., Diatoms: Self assembled silica nanostructures, and templates for bio/chemical sensors and biomimetic membranes. *Analyst*, 136, 42, 2011.
5. Yang, Y., Song, X., Li, X., Chen, Z., Zhou, C., Zhou, Q., Chen, Y., Recent progress in biomimetic additive manufacturing technology: From materials to functional structures. *Adv. Mater.*, 30, 1706539, 2018.
6. Du Plessis, A., Broeckhoven, C., Yadroitsava, I., Yadroitsev, I., Hands, C.H., Kunju, R., Bhate, D., Beautiful and functional: A review of biomimetic design in additive manufacturing. *Addit. Manuf.*, 27, 408, 2019.
7. Oyen, M., Nanoindentation of biological and biomimetic materials. *Exp. Tech.*, 37, 73–87, 2013.
8. Du Plessis, A. and Broeckhoven, C., Looking deep into nature: A review of micro-computed tomography in biomimicry. *Acta Biomater.*, 85, 27, 2019.
9. Rodríguez, R.E., Agarwal, S.P., An, S., Kazyak, E., Das, D., Shang, W., Dasgupta, N.P., Biotemplated Morpho butterfly wings for tunable structurally colored photocatalysts. *ACS Appl. Mater. Interfaces*, 10, 4614, 2018.
10. Zampieri, A., Mabande, G.T., Selvam, T., Schwieger, W., Rudolph, A., Hermann, R., Greil, P., Biotemplating of Luffa cylindrica sponges to self-supporting hierarchical zeolite macrostructures for bio-inspired structured catalytic reactors. *Mater. Sci. Eng. C*, 26, 130, 2006.
11. Paris, O., Burgert, I., Fratzl, P., Biomimetics and biotemplating of natural materials. *MRS Bull.*, 35, 219, 2010.
12. Katiyar, N.K., Goel, G., Hawi, S., Goel, S., Nature-inspired materials: Emerging trends and prospects. *NPG Asia Mater.*, 13, 1, 2021.
13. Kasonde, M. and Kanyanta, V., Future of superhard material design, processing and manufacturing, in: *Microstructure-Property Correlations for Hard, Superhard, and Ultrahard Materials*, pp. 211–239, Springer, Cham, 2016.
14. Ritchie, R.O., The conflicts between strength and toughness. *Nat. Mater.*, 10, 817, 2011.
15. Mayer, G., Trejo, R., Lara-Curzio, E., Rodriguez, M., Tran, K., Song, H., Ma, W.H., Lessons for new classes of inorganic/organic composites from the spicules and skeleton of the sea sponge *Euplectella aspergillum*. *MRS Online Proc. Lib. (OPL)*, 844, Y4.2.1, 2004.
16. Sarikaya, M., Fong, H., Sunderland, N., Flinn, B.D., Mayer, G., Mescher, A., Gaino, E., Biomimetic model of a sponge-spicular optical fiber—Mechanical properties and structure. *J. Mater. Res.*, 16, 1420, 2001.

17. Ghazlan, A., Ngo, T., Tan, P., Xie, Y.M., Tran, P., Donough, M., Inspiration from Nature's body armours–A review of biological and bioinspired composites. *Compos. Part B*, 205, 108513, 2021.
18. Hassan, S.A., Santulli, C., Yahya, M.Y., Gang, C.L., Abu Bakar, M.N.F.B., The potential of biomimetics design in the development of impact resistant material. *FME Trans.*, 46, 108, 2018.
19. Van Le, T., Ghazlan, A., Ngo, T., Nguyen, T., Performance of a biomimetic 3D printed conch-like structure under quasi-static loading. *Compos. Struct.*, 246, 112433, 2020.
20. Wang, X., Xiang, H., Sun, X., Liu, J., Hou, F., Zhou, Y., Mechanical properties and damage tolerance of bulk Yb3Al5O12 ceramic. *J. Mater. Sci. Technol.*, 31, 369, 2015.
21. Fleischli, F.D., Dietiker, M., Borgia, C., Spolenak, R., The influence of internal length scales on mechanical properties in natural nanocomposites: A comparative study on inner layers of seashells. *Acta Biomater.*, 4, 1694, 2008.
22. Menig, R., Meyers, M.H., Meyers, M.A., Vecchio, K.S., Quasi-static and dynamic mechanical response of Haliotis rufescens (abalone) shells. *Acta Mater.*, 48, 2383, 2000.
23. Perera, A.S. and Coppens, M.O., Re-designing materials for biomedical applications: From biomimicry to nature-inspired chemical engineering. *Philos. Trans. R. Soc. A*, 377, 20180268, 2019.
24. Alghamdi, S., Du, F., Yang, J., Pinder, G., Tan, T., Tensile and shear behavior of microscale growth layers between nacre in red abalone. *J. Mech. Phys. Solids*, 138, 103928, 2020.
25. Ko, K., Jin, S., Lee, S.E., Hong, J.W., Impact resistance of nacre-like composites diversely patterned by 3D printing. *Compos. Struct.*, 238, 111951, 2020.
26. Finnemore, A., Cunha, P., Shean, T., Vignolini, S., Guldin, S., Oyen, M., Steiner, U., Biomimetic layer-by-layer assembly of artificial nacre. *Nat. Commun.*, 3, 1, 2012.
27. Zhang, Y., Gong, S., Zhang, Q., Ming, P., Wan, S., Peng, J., Cheng, Q., Graphene-based artificial nacre nanocomposites. *Chem. Soc. Rev.*, 45, 2378, 2016.
28. Gil-Bona, A. and Bidlack, F.B., Tooth enamel and its dynamic protein matrix. *Int. J. Mol. Sci.*, 21, 4458, 2020.
29. Uskoković, V., Prospects and pits on the path of biomimetics: The case of tooth enamel. *J. Biomim. Biomater. Tissue Eng.*, 8, 45, 2010.
30. Uskoković, V., Biomineralization and biomimicry of tooth enamel, in: *Non-Metallic Biomaterials for Tooth Repair and Replacement*, P. Vallittu, (Ed.), p. 20, Woodhead Publishing Series in Biomaterials, Non-Metallic Biomaterials for Tooth Repair and Replacement, Woodhead Publishing, Sawston, Cambridge, UK, 2013.
31. Pandya, M. and Diekwisch, T.G., Enamel biomimetics—Fiction or future of dentistry. *Int. J. Oral. Sci.*, 11, 1, 2019.

32. Wu, D., Yang, J., Li, J., Chen, L., Tang, B., Chen, X., Li, J., Hydroxyapatite-anchored dendrimer for in situ remineralization of human tooth enamel. *Biomater.*, 34, 5036, 2013.
33. Eadie, L. and Ghosh, T.K., Biomimicry in textiles: Past, present and potential. An overview. *J. R. Soc. Interface*, 8, 761, 2011.
34. Ellison, M.S., Biomimetic textiles, in: *Bioinspiration, Biomimetics, and Bioreplication*, p. 8686, International Society for Optics and Photonics, 868602, Bellingham, Washington, USA, 2013.
35. Riekel, C. and Vollrath, F., Spider silk fibre extrusion: Combined wide-and small-angle X-ray microdiffraction experiments. *Int. J. Biol. Macromol.*, 29, 203, 2001.
36. Chen, X., Shao, Z., Vollrath, F., The spinning processes for spider silk. *Soft Matter*, 2, 448, 2006.
37. Waite, J.H., Qin, X.X., Coyne, K.J., The peculiar collagens of mussel byssus. *Matrix Biol.*, 17, 93, 1998.
38. Porter, D. and Vollrath, F., Silk as a biomimetic ideal for structural polymers. *Adv. Mater.*, 21, 487, 2009.
39. Rising, A., Widhe, M., Johansson, J., Hedhammar, M., Spider silk proteins: Recent advances in recombinant production, structure–function relationships and biomedical applications. *Cell Mol. Life Sci.*, 68, 169, 2011.
40. Heidebrecht, A., Eisoldt, L., Diehl, J., Schmidt, A., Geffers, M., Lang, G., Scheibel, T., Biomimetic fibers made of recombinant spidroins with the same toughness as natural spider silk. *Adv. Mater.*, 27, 2189, 2015.
41. Rawat, P., Zhu, D., Rahman, M.Z., Barthelat, F., Structural and mechanical properties of fish scales for the bio-inspired design of flexible body armors: A review. *Acta Biomater.*, 121, 41, 2021.
42. Lin, Y.S., Wei, C.T., Olevsky, E.A., Meyers, M.A., Mechanical properties and the laminate structure of Arapaima gigas scales. *J. Mech. Behav. Biomed. Mater.*, 4, 7, 1145, 2011.
43. Suzuki, A., Kodama, Y., Miwa, K., Kishimoto, K., Hoshikawa, E., Haga, K., Izumi, K., Manufacturing micropatterned collagen scaffolds with chemical-crosslinking for development of biomimetic tissue-engineered oral mucosa. *Sci. Rep.*, 10, 1, 2020.
44. Feng, H., Li, X., Deng, X., Li, X., Guo, J., Ma, K., Jiang, B., The lamellar structure and biomimetic properties of a fish scale matrix. *RSC Adv.*, 10, 875, 2020.
45. Bianciardi, A., Credi, C., Levi, M., Rosa, F., Zecca, A., Biomimicry thinking: Methodological improvements and practical implementation. *Bioinspired Biomim. Nanobiomaterials*, 6, 87, 2017.
46. Santulli, C., Interaction with water in nature and self-cleaning potential of biological materials and species. *Cuad. Cent. Estud. Diseñ. Comun.*, 22, 151, 2019.
47. Santulli, C., The design of plant fibres as biomimetic composites. *Cuad. Cent. Estud. Diseñ. Comun.*, 149, 131, 2021.

48. Lin, J., Cai, Y., Wang, X., Ding, B., Yu, J., Wang, M., Fabrication of biomimetic superhydrophobic surfaces inspired by lotus leaf and silver ragwort leaf. *Nanoscale*, *3*, 1258, 2011.
49. Zorba, V., Stratakis, E., Barberoglou, M., Spanakis, E., Tzanetakis, P., Anastasiadis, S.H., Fotakis, C., Biomimetic artificial surfaces quantitatively reproduce the water repellency of a lotus leaf. *Adv. Mater.*, 20, 4049, 2008.
50. Rather, A.M., Shome, A., Kumar, S., Bhunia, B.K., Mandal, B.B., Srivastava, H.K., Manna, U., Alkali metal-ion assisted Michael addition reaction in controlled tailoring of topography in a superhydrophobic polymeric monolith. *J. Mater. Chem. A*, 6, 17019, 2018.
51. Higham, T.E., Russell, A.P., Niewiarowski, P.H., Wright, A., Speck, T., The ecomechanics of gecko adhesion: Natural surface topography, evolution, and biomimetics. *Integr. Comp. Biol.*, 59, 148, 2019.
52. Mahdavi, A., Ferreira, L., Sundback, C., Nichol, J.W., Chan, E.P., Carter, D.J., Karp, J.M., A biodegradable and biocompatible gecko-inspired tissue adhesive. *Proc. Natl. Acad. Sci.*, 105, 2307, 2008.
53. Rai, R., Tallawi, M., Grigore, A., Boccaccini, A.R., Synthesis, properties and biomedical applications of poly (glycerol sebacate) (PGS): A review. *Prog. Polym. Sci.*, 37, 1051, 2012.
54. Wang, L., Hui, Y., Fu, C., Wang, Z., Zhang, M., Zhang, T., Recent advances in Gecko-inspired adhesive materials and application. *J. Adhes. Sci. Technol.*, 34, 2275, 2020.
55. Bandara, N., Zeng, H., Wu, J., Marine mussel adhesion: Biochemistry, mechanisms, and biomimetics. *J. Adhes. Sci. Technol.*, 27, 2139, 2013.
56. Vreeland, V., Waite, J.H., Epstein, L., Minireview—Polyphenols and oxidases in substratum adhesion by marine algae and mussels. *J. Phycol.*, 34, 1, 1998.
57. Numata, K. and Baker, P.J., Synthesis of adhesive peptides similar to those found in blue mussel (Mytilus edulis) using papain and tyrosinase. *Biomacromol.*, 15, 3206, 2014.
58. Danner, E.W., Kan, Y., Hammer, M.U., Israelachvili, J.N., Waite, J.H., Adhesion of mussel foot protein Mefp-5 to mica: An underwater superglue. *Biochem.*, 51, 6511, 2012.
59. Lee, C., Lee, G.W., Choi, W., Yoo, C.H., Chun, B., Lee, J.S., Jung, H.W., Pattern flow dynamics over rectangular Sharklet patterned membrane surfaces. *Appl. Surf. Sci.*, 514, 2020, 145961.
60. Reddy, S.T., Chung, K.K., McDaniel, C.J., Darouiche, R.O., Landman, J., Brennan, A.B., Micropatterned surfaces for reducing the risk of catheter-associated urinary tract infection: An *in vitro* study on the effect of sharklet micropatterned surfaces to inhibit bacterial colonization and migration of uropathogenic Escherichia coli. *J. Endourol.*, 25, 1547, 2011.
61. Wen, L., Weaver, J.C., Lauder, G.V., Biomimetic shark skin: Design, fabrication and hydrodynamic function. *J. Exp. Biol.*, 217, 1656, 2014.
62. Huang, J., Wang, Q., Wu, Z., Ma, Z., Yan, C., Shi, Y., Su, B., 3D-printed underwater super-oleophobic shark skin toward the electricity generation through

low-adhesion sliding of magnetic nanofluid droplets. *Adv. Funct. Mater.*, 31, 2103776, 2021.
63. Padole, M., Gharde, S., Kandasubramanian, B., Three-dimensional printing of molluscan shell inspired architectures via fused deposition modeling. *Envir. Sci. Pollut. Res.*, 28, 46356–46366, 2021.
64. Zhao, H., Yang, Z., Guo, L., Nacre-inspired composites with different macroscopic dimensions: Strategies for improved mechanical performance and applications. *NPG Asia Mater.*, 10, 4, 1–22, 2018.
65. Wijerathne, B., Liao, T., Ostrikov, K., Sun, Z., Bioinspired robust mechanical properties for advanced materials. *Small Struct.*, 2100228, 2022.
66. Behera, A., Self-healing materials, in: *Advanced Materials*, pp. 321–358, Springer, Cham, 2022.
67. Korde, J.M. and Kandasubramanian, B., Naturally biomimicked smart shape memory hydrogels for biomedical functions. *Chem. Eng. J.*, 379, 122430, 2020.
68. Knapczyk-Korczak, J. and Stachewicz, U., Biomimicking spider webs for effective fog water harvesting with electrospun polymer fibers. *Nanoscale*, 13, 16034, 2021.
69. Deo, D., Singh, S.P., Mohanty, S., Guhathakurata, S., Pal, D., Mallik, S., Biomimicking of phyto-based super-hydrophobic surfaces towards prospective applications: A review. *J. Mater. Sci.*, 57, 1–28, 2022.
70. Tian, Y., Liu, Y., Su, Z., Wang, S., Zhang, B., Zhang, H., Zhang, Q., Biomimetic brushlike slippery coatings with mechanically robust, self-cleaning, and icephobic properties. *ACS Appl. Mater. Interfaces*, 12, 48, 54041–54052, 2020.
71. Yang, Y., Shen, H., Qiu, J., Fabrication of biomimetic robust self-cleaning superhydrophobic wood with canna-leaf-like micro/nanostructure through morph-genetic method improved water-, UV-, and corrosion resistance properties. *J. Mol. Struct.*, 1219, 128616, 2020.

9

Novel Biomimicry Techniques for Detecting Plant Diseases

Adeshina Fadeyibi[1]* and Mary Fadeyibi[2]

[1]Department of Food and Agricultural Engineering, Faculty of Engineering and Technology, Kwara State University, Ilorin, Kwara State, Nigeria
[2]Department of Biological Science, Faculty of Science, Augustine University, Ilara-Epe, Lagos State, Nigeria

Abstract
Biomimicry techniques (BmT) can be regarded as crop management concepts deployed for detecting, measuring, and responding to diseases during agricultural production. In this paper, various approaches of the BmT for the efficient management of unhealthy plants during preharvest and postharvest operations were reviewed. The remote sensors, machine vision, and robotics BmT approaches are mostly applied during preharvest operations for detecting plant diseases. Certain diseases, such as white-soft, gray-mold, and Sclerotium rots, may attack vegetables after harvest, thus requiring a high resolution BmT, like neural network and support vector machine to manage. To the best of our knowledge, this technology is mostly applied for protecting tropical and temperate fruits and vegetables but have not been tested for cereal and legume plants disease management. Also, the resulting image from the BmT can sometimes be misleading due to bad weather and other factors related to equipment performance. Thus, this study will help provide meaningful insights for the advancement of the technology on plant management practice during preharvest and postharvest processing and storage.

Keywords: Plant disease, detection, preharvest and postharvest management, biomimicry techniques

**Corresponding author*: deshina.fadeyibi@kwasu.edu.ng

Inamuddin, Tariq Altalhi and Ashjan Alrogi (eds.) Biomimicry Materials and Applications, (199–214)

9.1 Introduction

Plant physiology can be influenced by environmental factors such as temperature, rainfall, and soil condition [1]. It can also be affected by diseases from bacteria, yeast, fungi, and other pathogens [2–5]. Sometimes, ectoparasites, like insects, mites, vertebrate, and other pests, can affect plant health by eating its tissue. The physical damage imposed by these agents can be so enormous if adequate preventive measures are not taken before harvest and during processing or storage [6, 7]. Thus, the control of the plant diseases is crucial to the production of food that can meet the population need, and provide a significant solution to the agricultural land, and water uses [8]. However, most crops can have *in vitro* ability for the disease control without involving any external agent [9, 10]. The plants can be genetically engineered to possess high resistance abilities to protect themselves against external agents that can cause diseases [11, 12]. Thus, continuous advances in the science of plant pathology are essential to improve disease control [13, 14], and ensure healthy food products availability during preharvest and postharvest operations [15–18]. Structurally, this category of the plant contains layers of stomata guard, cuticle, and cell wall that can act on the pathogens and render them inactive by creating a strong intercellular barrier against the pathogens [5, 19, 20].

A physical observation or visual monitoring of the plant on and off the farm may not effectively identify the disease or prevent it. There is a need therefore for an improved and effective technology that can provide in-depth feasibility of the nature of the disease and facilitate plant protection during processing and storage [7, 21]. Many different techniques have been described for detecting plant diseases before and after harvest operations in agricultural production [22, 23]. The remote sensing, machine vision and robotics are among the most applied BmT for the detection of the physiological changes in the plant occurring due to the disease infestation. Certain diseases, like white-soft, gray-mold and Sclerotium rots, may attack vegetables after harvest, thus requiring a high resolution BmT, like neural network and support vector machine to manage. However, these require high level of technical know-how to operate and are expensive thereby making them less attractive to the farmers. Also, it is mostly applied for protecting tropical and temperate fruits and vegetables that are used either as spices or for medicinal purposes. This review provides detailed information on how these techniques can effectively act on the plant to detect the diseases and provide protection. We also proposed the use of the BmTs for protecting cereal and legume plants during preharvest and postharvest processing and storage.

9.2 Preharvest Biomimicry Detection Techniques

9.2.1 Remote Sensing Technique Approach

This technique permits the detection and protection of the physical behavior or characteristics of the plant by measuring the way it reflects and emits radiation at a distance or before harvest [24–26]. The remote sensor (RS) can also enable a quantitative and qualitative assessment of the plant growth indices such leave area, stem size, root penetration, and son on, thus revealing the health status of the plant [1, 27–29], as shown in Figure 9.1 (a) and (b) for single and multiphase RS imaging, respectively. According to Zhang *et al.* [30], the RS is a useful technique for monitoring diseases at every growth cycle stage of the plant system. In another related study, Feng *et al.* [31] proposed the use of unmanned aerial vehicle RS technique to monitor the plant growth and physiology on the farm especially the influence of genotypes and environment on plant growth. Also, Borges *et al.* [32] designed a pocket RS for controlling and quantifying the color of plant leave. The technique was able to sense the nutritional and physiological status of the plant. The RS tool has also been successfully applied for the detection of diseases in banana plant, classifying color changes due to the disease infestation [33] and for rapid screening of a death syndrome diseases in soybeans [29]. A high success rate was reported for detecting the banana bunchy top and Xanthomonas Wilt of Banana diseases. In precision

Figure 9.1 A typical map of plant diseases with a remote sensing technique [30].

agriculture, the RS has also found practical application [34]. Thanks to the high throughput of this practice, a site-specific fungicide may be applied to infested areas if the disease is stable, although the traditional uniform application is more suitable for diseases that can spread quickly through the field. Although a wide range of practical applications of the RS for plant disease monitoring and control have been reported, the commercial application of this technology is not yet exploited. Also, the RS tool requires a high-level technical ability to operate, thus the uneducated farmers may not be able to handle it effectively. Sometimes the image obtained may not reflects the plant disease scenario due to bad weather and other factors that can affect the equipment performance. There is, therefore, the need to simplify this technology to make it less technical and accommodate different weather variables for wider preharvest plant disease detection applications.

9.2.2 Machine Vision and Fuzzy Logic Approaches

Machine vision (MV) and fuzzy logic (FL) are novel biomimicry techniques applied for detecting plant diseases during preharvest processing. The MV is an enabling new automated process in agriculture, which promotes greater efficiency [35–37], while the fuzzy logic (FL) can capture the required knowledge of the agriculture domain and detect or control, for example, the plant diseases during preharvest operations. These two techniques can be integrated into the precision agriculture technology process to ensure an effective automation of the agricultural processes including the detection, monitoring, and control of the various degree of the plant that may occur during its growth cycle [38–40]. A novel MV was proposed for the identification and classification of the plant leaf diseases by Dhingra *et al.* [41]. The quality of the leaf defines the degree of excellence or a state of being free from defects, deficits, or variations [42]. According to Iqbal *et al.* [43], who reviewed the potentials of the image processing technology for detecting diseases in citrus plant, the MV is an excellent tool for detecting citrus plant diseases, such as canker, blackspot, scab, melanose, greening, and anthracnose defects. A typical citrus plant diseases detected using the MV technique is shown in Figure 9.2. If not corrected, this can pose serious threats to the economic and production status of the plant [44]. Thus, the MV process can help decrease the over dependence on human labor for protection offered to the growing plant [43, 45–47]. Hence, the planning activities on the farm, which includes routine fertilizer application and irrigation, will be encouraged. Meanwhile, the application of FL tool in plant disease detection is relatively new. So far, it has only been applied for the automatic detection of the black measles diseases [48] in

Figure 9.2 Typical citrus plant diseases detected using the MV technique [43].

grapes and for wolfberry pickling [49]. To best of our knowledge, the MV and FL technique has not been applied for detecting preharvest diseases in cereal and legume plants. Thus, these present a new frontier for future research.

9.2.3 Robotics Approach

Robotics (RO) is an artificial intelligent system designed to perform human tasks with greater accuracy and efficiency [50, 51]. In agriculture, the RO is used to automate slow, dull, and repeating tasks on the field, thus allowing the farmers to focus more on the improvement of plant yields [52, 53]. A typical robotic system for automatic in-field leaf disease detection is shown in Figure 9.3. A pneumatic controlled robotic system, which was developed by Gao *et al.* [54] for harvesting or picking cherry tomatoes fruits on the farm, can be modified to assist in detecting the disease of the product before and during the harvest operations. To archive this, certain detection accessories like the hyperspectral MV must be attached to the robot, especially around the picking device. An experimental electric robotic platform, mainly destined for spraying operations, was developed by Loukatos *et al.* [55] to improve its functions and to host a high-quality optical camera for detecting and capturing areas of defects on the plant parts. Also, Chen *et al.* [52] designed an automated robotic system for in-field measurement of maize leaf diseases as shown in Figure 9.3. The technology uses hyperspectral imaging system to sense and detect areas of defects around the leaf and send responses to the farmer who will in turn provide timely treatment for

Figure 9.3 A robotic system for automatic in-field leaf disease detection [52].

the defective plant. However, the RO technology is highly technical to operate and costly, in terms of equipment and energy costs. Thus, it is less attractive, and many peasant farmers may not be able to afford it. Hence, research is needed to build a less technical and more affordable RO system for use in the detection of the plant diseases on the farm.

9.3 Postharvest Biomimicry Detection Techniques

9.3.1 Neural Network Approach

Neural network (NN) provides a mapping between the image of the disease plant (input) to the crop disease pair (output) [56, 57], as shown in Figure 9.4. This involves creating a network structure that can effectively connect the nodes, edge-weights, input, and output variables together [58, 59]. The network parameters can be turned to ensure adequate training of the input data, as characterized in Table 9.1. To develop accurate algorism for the purpose of the plant disease diagnosis, images of diseased plants were normally compared with those of the healthy plants [60–63]. The application of the NN for detecting various diseases in various plant parts has been widely studied and reported in the literature [64–69]. Certain diseases, like white-soft, gray-mold and Sclerotium rots, may attack vegetables after harvest, thus requiring a high resolution BmT, like the SVM to manage. For example, Zhang *et al.* [70] developed the NN tool for the smart disease detection in apple using an Internet of Things (IoT) device to extract the color and texture features of the food product. This NN could combine the features that are extracted by color and texture analysis

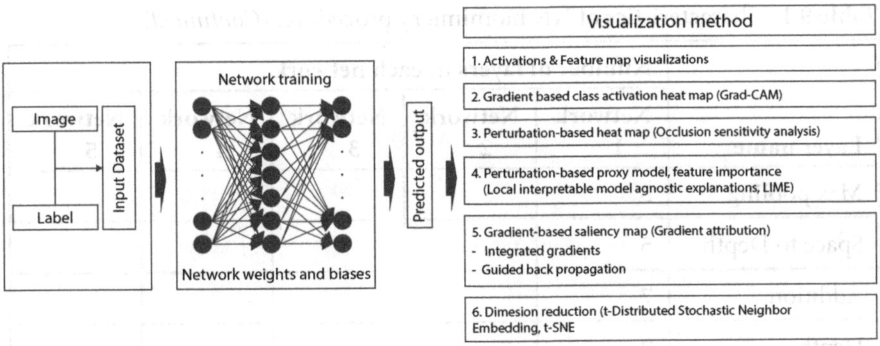

Figure 9.4 Plant disease classification procedure using NN biomimicry technique [73].

Table 9.1 Characteristics of NN biomimicry procedure.

	Number of layers in each network				
Layer name	**Network 1**	**Network 2**	**Network 3**	**Network 4**	**Network 5**
Image input	1				
Convolution	106				
Group convolution	7				
Batch normalization	105	102			100
Instance normalization	1	4			6
Group normalization	7				
Re LU	-	-	20	-	-
Leaky	-	20	-	-	-
Swish	108	88	88	88	87
Elu	-	-	-	20	-
Clipped Re LU	-	-	-	-	21
Avg Pooling	13				

(Continued)

Table 9.1 Characteristics of NN biomimicry procedure. (*Continued*)

Layer name	Number of layers in each network				
	Network 1	Network 2	Network 3	Network 4	Network 5
Max pooling	3				
Space to Depth	5				
Addition	7				
Depth concatenation	9				
Dropout	5				
Fully connected	1				
Soft max	1				
Class output	1				

Source: Islam *et al.* [73].

subnets, thereby improving the resulting disease image expression of the infected apple. The NN has also been applied as a smart device for the provision of automatic diagnosis of corn diseases and prevent severe losses in silos and bins [71, 72]. This technology was also effective for detecting corn leaf disease with an accuracy of 88.46%. Also, Islam *et al.* [73] and Agarwal *et al.* [74] visualized the tomato leaf disease by combining several convolutional NN approaches and predicted the performance of each NN protocol in response to their abilities to recognize the plant diseases, as shown in Figure 9.4. Regardless of the type of approach used, the authors concluded that the NN provided overall training, learning, and detection accuracy through the ability to visualize disease symptoms on the target tomato leaf. However, as promising as the NN technique may sound, it can encounter greater computational problem and is highly dependent on the hardware system. The network may malfunction unexpectedly thus making prediction unreliable at times. Therefore, further research is needed to address these challenges.

9.3.2 Support Vector Machine Approach

Support vector machine (SVM) is a supervised learning model that can analyze data for classification and recognition purpose [75, 76]. Recently, the SVM has been applied in agriculture, alone or in combination with other techniques, to detect and classify different types of plant leaf diseases. For example, a combination of RS and SVM technologies was reported by Griffel *et al.* [77] and Mokhtar *et al.* [78] for detecting and differentiating the diseases as they occur during postharvest processing of potato and tomato crops. This technique to detect viral diseases in potato (Figure 9.5), leading to an improvement in the quality of the product needed to maximize yields. Rumpf *et al.* [79] reported a procedure for early detection and classification of diseases in sugar beet based on the application of the SVM and spectral vegetation indices. The technique was able to differentiate and identify the various diseases, including cercospora leaf spot, powdery mildew diseases, before their symptoms are revealed. Also, Leena and Saju [80] reported another SVM technology was effective for classifying the micronutrient-deficient parts of a harvested maize crop. The potential of radial basis SVM and NN approaches were compared by Omrani *et al.* [81] for detecting apple diseases. The authors concluded that the SVM was a better tool especially for identifying and classifying apple black spot, *Alternaria*, and apple leaf miner diseases of the harvested apple fruit. To the best of our knowledge, there is not enough information on the commercial applications of SVM technique for detecting and classifying diseases of cereals and legume plants. Thus, this presents a new area for future studies.

Figure 9.5 Potato plant (a) appearing healthy before harvest (b) exhibiting strong visual symptoms of potato virus disease with stunted growth after harvest as detected using the SVM technique [77].

9.4 Prospects and Conclusion

A visual control and monitoring of the plant growth during preharvest and postharvest processing and storage may not effectively identify diseases or provide protection for future use. In this paper, several novel procedures for plant disease detection including the RS, NN, MV, RO, and SVM, which are noted for their unique identification abilities were reviewed. Although a wide range of practical applications of the RS for plant disease monitoring and control have been reported, the commercial application of this technology is not yet exploited. It also requires a high-level technical ability to operate, thus the uneducated farmers may not be able to handle it effectively. Recent studies have shown that the MV technique can lower the over dependence on human labor for plant protection. It was used to detect and quantify the level of nutritional properties in the plant at every stage of its growth cycle. However, the application the FL tool in plant disease detection is relatively new. So far, it has only been applied for the automatic detection of the black measles diseases in grapes and for wolfberry pickling. The RO technology is highly technical to operate and costly, in terms of equipment and energy costs. Thus, it is less attractive, and many peasant farmers may not be able to afford it. Hence, research is needed to build a less technical and more affordable RO system for use in the detection of the plant diseases on the farm. Also, to best of our knowledge, the MV and FL technique has not been applied for detecting diseases in cereals and legumes plants. In general, the image obtained from these tools may sometimes not reflects the actual plant disease scenario due to bad weather and other factors related to equipment performance. There is therefore the need to simplify these technologies to make them less technical and accommodate different weather variables for wider applications. These present new frontiers for future research.

References

1. Ali, M.M., Bachik, N.A., Muhadi, N.A., Tuan Yusof, T.N., Gomes, C., Non-destructive techniques of detecting plant diseases: A review. *Physiol. Mol. Plant Pathol.*, 108, 1–10, 2019.
2. Zhang, J., Huang, Y., Pu, R., Gonzalez-Moreno, P., Yuan, L., Wu, K., Huang, W., Monitoring plant diseases and pests through remote sensing technology: A review. *Comput. Electron. Agric.*, 165, 104943, 2019.

3. Bischoff, V., Farias, K., Menzen, J.P., Pessin, G., Technological support for detection and prediction of plant diseases: A systematic mapping study. *Comput. Electron. Agric.*, 181, 105922, 2021.
4. Sanghavi, K., Sanghavi, M., Rajurkar, A.M., Early stage detection of downey and powdery mildew grape disease using atmospheric parameters through sensor nodes. *Artif. Intell. Agric.*, 5, 223–232, 2021.
5. Amarasingam, N., Ashan Salgadoe, A.S., Powell, K., Gonzalez, L.F., Natarajan, S., A review of UAV platforms, sensors, and applications for monitoring of sugarcane crops. *Remote Sens. Appl.: Soc. Environ.*, 26, 100712, 2022.
6. Fadeyibi, A., Osunde, Z., Applied, M.Y., Effects of period and temperature on quality and shelf-life of cucumber and garden-eggs packaged using cassava starch-zinc nanocomposite film. *J. Appl. Packag. Res.*, 12, 1, 1–10, 2020.
7. Fadeyibi, A., Lamidi, W., Ademola, S., Engineering and proximate properties of miracle berry fruit (Synsepalum dulcificum L.) essential for its processing equipment design. *Agric. Eng. Int.: CIGR J.*, 23, 4, 227–235, 2021.
8. Mohammad-Razdari, A., Rousseau, D., Bakhshipour, A., Taylor, S., Poveda, J., Kiani, H., Recent advances in E-monitoring of plant diseases. *Biosens. Bioelectron.*, 201, 113953, 2022.
9. Yousfi, S., Marín, J., Parra, L., Lloret, J., Mauri, P.V., Remote sensing devices as key methods in the advanced turfgrass phenotyping under different water regimes. *Agric. Water Manage.*, 266, 107581, 2022.
10. Jafarbiglu, H. and Pourreza, A., A comprehensive review of remote sensing platforms, sensors, and applications in nut crops. *Comput. Electron. Agric.*, 197, 106844, 2022.
11. Zhou, Y., Flynn, K.C., Gowda, P.H., Wagle, P., Ma, S., Kakani, V.G., Steiner, J.L., The potential of active and passive remote sensing to detect frequent harvesting of alfalfa. *Int. J. Appl. Earth Obs. Geoinf.*, 104, 102539, 2021.
12. He, Y., Chen, G., Potter, C., Meentemeyer, R.K., Integrating multi-sensor remote sensing and species distribution modeling to map the spread of emerging forest disease and tree mortality. *Remote Sens. Environ.*, 231, 111238, 2019.
13. Jiao, W., Wang, L., McCabe, M.F., Multi-sensor remote sensing for drought characterization: Current status, opportunities and a roadmap for the future. *Remote Sens. Environ.*, 256, 112313, 2021.
14. Cardoso, R.M., Pereira, T.S., Facure, M.H.M., dos Santos, D.M., Mercante, L.A., Mattoso, L.H.C., Correa, D.S., Current progress in plant pathogen detection enabled by nanomaterials-based (bio)sensors. *Sens. Actuators Rep.*, 4, 100068, 2022.
15. Fadeyibi, A., Osunde, Z.D., Yisa, M.G., Okunola, A.A., Investigation into properties of starch-based nanocomposite materials for fruits and vegetables packaging-A review. *FUTA J. Eng. Technol.*, 11, 1, 12–16, 2017.
16. Alabi, K., Olalusi, A., Olaniyan, A., Fadeyibi, A., Gabriel, L., Effects of osmotic dehydration pretreatment on freezing characteristics and quality of frozen fruits and vegetables. *J. Food Process Eng.*, 48, 8, 14037, 2022.

17. Fadeyibi, A., Osunde, Z.D., Yisa, M.G., Prediction of some physical attributes of cassava starch–zinc nanocomposite film for food-packaging applications. *J. Package Technol. Res.*, 3, 1, 35–41, 2019.
18. Wang, T., Mei, X., Alex Thomasson, J., Yang, C., Han, X., Yadav, P.K., Shi, Y., GIS-based volunteer cotton habitat prediction and plant-level detection with UAV remote sensing. *Comput. Electron. Agric.*, 193, 106629, 2022.
19. Lassalle, G., Monitoring natural and anthropogenic plant stressors by hyperspectral remote sensing: Recommendations and guidelines based on a meta-review. *Sci. Total Environ.*, 788, 147758, 2021.
20. Xie, C. and Yang, C., A review on plant high-throughput phenotyping traits using UAV-based sensors. *Comput. Electron. Agric.*, 178, 105731, 2020.
21. Fadeyibi, A., Modeling rheological behavior of beef based on time-dependent deformation and packaging. *Gazi Univ. J. Sci.*, 35, 3, 997–1008, 2022.
22. Fadeyibi, A., Osunde, Z.D., Yisa, M.G., Optimization of processing parameters of nanocomposite film for fresh sliced okra packaging. *J. Appl. Packag. Res.*, 11, 2, 1–10, 2019.
23. Fadeyibi, A., Osunde, Z.D., Egwim, E.C., Idah, P.A., Performance evaluation of cassava starch-zinc nanocomposite film for tomatoes packaging. *J. Agric. Eng.*, 48, 3, 137–146, 2017.
24. Lin, C., Labzovskii, L.D., Mak, H.W.L., Fung, J.C.H., Lau, A.K.H., Kenea, S.T., Bilal, M., vande Hey, J.D., Lu, X., Ma, J., Observation of PM2.5 using a combination of satellite remote sensing and low-cost sensor network in Siberian urban areas with limited reference monitoring. *Atmos. Environ.*, 227, 117410, 2020.
25. Žibrat, U., Širca, S., Susič, N., Knapič, M., Gerič Stare, B., Urek, G., Noninvasive detection of plant parasitic nematodes using hyperspectral and other remote sensing systems, in: *Hyperspec Rem Sen: Theory and Applications*, pp. 357–375, 2020.
26. Kattenborn, T., Leitloff, J., Schiefer, F., Hinz, S., Review on convolutional neural networks (CNN) in vegetation remote sensing. *ISPRS J. Photogramm. Remote Sens.*, 173, 24–49, 2021.
27. Kabisch, N., Selsam, P., Kirsten, T., Lausch, A., Bumberger, J., A multi-sensor and multi-temporal remote sensing approach to detect land cover change dynamics in heterogeneous urban landscapes. *Ecol. Indic.*, 99, 273–282, 2019.
28. Lin, F., Weng, Y., Chen, H., Zhuang, P., Intelligent greenhouse system based on remote sensing images and machine learning promotes the efficiency of agricultural economic growth. *Environ. Technol. Innov.*, 24, 101758, 2021.
29. Hatton, N., Sharda, A., Schapaugh, W., van der Merwe, D., Remote thermal infrared imaging for rapid screening of sudden death syndrome in soybean. *Comput. Electron. Agric.*, 178, 1–10, 2020.
30. Zhang, J., Huang, Y., Pu, R., Gonzalez-Moreno, P., Yuan, L., Wu, K., Huang, W., Monitoring plant diseases and pests through remote sensing technology: A review. *Comput. Electron. Agric.*, 165, 104943, 2019.

31. Feng, L., Chen, S., Zhang, C., Zhang, Y., He, Y., A comprehensive review on recent applications of unmanned aerial vehicle remote sensing with various sensors for high-throughput plant phenotyping. *Comput. Electron. Agric.*, 182, 106033, 2021.
32. Borges, C.S., Vega R, R.A., Chakraborty, S., Weindorf, D.C., Lopes, G., Guimarães Guilherme, L.R., Curi, N., Li, B., Ribeiro, B.T., Pocket-sized sensor for controlled, quantitative and instantaneous color acquisition of plant leaves. *J. Plant Physiol.*, 272, 153686, 2022.
33. Gomez Selvaraj, M., Vergara, A., Montenegro, F., Alonso Ruiz, H., Safari, N., Raymaekers, D., Ocimati, W., Ntamwira, J., Tits, L., Omondi, A.B., Blomme, G., Detection of banana plants and their major diseases through aerial images and machine learning methods: A case study in DR Congo and Republic of Benin. *ISPRS J. Photogramm. Remote Sens.*, 169, 110–124, 2020.
34. Yang, C., Remote sensing and precision agriculture technologies for crop disease detection and management with a practical application example. *Eng.*, 6, 5, 528–532, 2020.
35. Singh, V. and Misra, A.K., Detection of plant leaf diseases using image segmentation and soft computing techniques. *Inf. Process. Agric.*, 4, 1, 41–49, 2017.
36. Sowmya, B.J., Shetty, C., Seema, S., Srinivasa, K.G., Utility system for premature plant disease detection using machine learning, in: *Hybrid Comput Intel*, pp. 149–172, 2020.
37. Jafari, M., Minaei, S., Safaie, N., Detection of pre-symptomatic rose powdery-mildew and gray-mold diseases based on thermal vision. *Infrared Phys. Technol.*, 85, 170–183, 2017.
38. Pane, C., Manganiello, G., Nicastro, N., Ortenzi, L., Pallottino, F., Cardi, T., Costa, C., Machine learning applied to canopy hyperspectral image data to support biological control of soil-borne fungal diseases in baby leaf vegetables. *Biol. Control*, 164, 104784, 2021.
39. Gladju, J., Kamalam, B.S., Kanagaraj, A., Applications of data mining and machine learning framework in aquaculture and fisheries: A review. *Smart Agric. Technol.*, 2, 100061, 2022.
40. Nturambirwe, J.F.I. and Opara, U.L., Machine learning applications to non-destructive defect detection in horticultural products. *Biosyst. Eng.*, 189, 60–83, 2020.
41. Dhingra, G., Kumar, V., Joshi, H.D., A novel computer vision based neutrosophic approach for leaf disease identification and classification. *Meas.: J. Int. Meas. Confed.*, 135, 782–794, 2019.
42. Xue, J., Zhang, L., Grift, T.E., Variable field-of-view machine vision based row guidance of an agricultural robot. *Comput. Electron. Agric.*, 84, 85–91, 2012.
43. Iqbal, Z., Khan, M.A., Sharif, M., Shah, J.H., ur Rehman, M.H., Javed, K., An automated detection and classification of citrus plant diseases using image processing techniques: A review. *Comput. Electron. Agric.*, 153, 12–32, 2018.

44. Kaur, N. and Devendran, V., Novel plant leaf disease detection based on optimize segmentation and law mask feature extraction with SVM classifier. *Mater Today: Proc.*, 2020.
45. Tian, Z., Ma, W., Yang, Q., Duan, F., Application status and challenges of machine vision in plant factory—A review. *Inf. Process. Agric.*, 9, 2, 195–211, 2021.
46. Sujatha, R., Chatterjee, J.M., Jhanjhi, N.Z., Brohi, S.N., Performance of deep learning vs machine learning in plant leaf disease detection. *Microprocess. Microsyst.*, 80, 103615, 2021.
47. Barbedo, J.G.A., Detection of nutrition deficiencies in plants using proximal images and machine learning: A review. *Comput. Electron. Agric.*, 162, 482–492, 2019.
48. Ji, M. and Wu, Z., Automatic detection and severity analysis of grape black measles disease based on deep learning and fuzzy logic. *Comput. Electron. Agric.*, 193, 106718, 2022.
49. Ma, Y., Zhang, W., Qureshi, W.S., Gao, C., Zhang, C., Li, W., Autonomous navigation for a wolfberry picking robot using visual cues and fuzzy control. *Inf. Process. Agric.*, 8, 1, 15–26, 2021.
50. Hejazipoor, H., Massah, J., Soryani, M., Asefpour Vakilian, K., Chegini, G., An intelligent spraying robot based on plant bulk volume. *Comput. Electron. Agric.*, 180, 105859, 2021.
51. Yoganandhan, A., Rajesh Kanna, G., Subhash, S.D., Hebinson Jothi, J., Retrospective and prospective application of robots and artificial intelligence in global pandemic and epidemic diseases. *Vacunas*, 22, 2, 98–105, 2021.
52. Chen, Z., Wang, J., Wang, T., Song, Z., Li, Y., Huang, Y., Wang, L., Jin, J., Automated in-field leaf-level hyperspectral imaging of corn plants using a Cartesian robotic platform. *Comput. Electron. Agric.*, 183, 105996, 2021.
53. Montoya-Cavero, L.E., Díaz de León Torres, R., Gómez-Espinosa, A., Escobedo Cabello, J.A., Vision systems for harvesting robots: Produce detection and localization. *Comput. Electron. Agric.*, 192, 106562, 2022.
54. Gao, J., Zhang, F., Zhang, J., Yuan, T., Yin, J., Guo, H., Yang, C., Development and evaluation of a pneumatic finger-like end-effector for cherry tomato harvesting robot in greenhouse. *Comput. Electron. Agric.*, 197, 106879, 2022.
55. Loukatos, D., Templalexis, C., Lentzou, D., Xanthopoulos, G., Arvanitis, K.G., Enhancing a flexible robotic spraying platform for distant plant inspection via high-quality thermal imagery data. *Comput. Electron. Agric.*, 190, 106462, 2021.
56. Ibrahim, M.H., Wba-Dnn: A hybrid weight bat algorithm with deep neural network for classification of poisonous and harmful wild plants. *Comput. Electron. Agric.*, 190, 1–10, 2021.
57. Golhani, K., Balasundram, S.K., Vadamalai, G., Pradhan, B., A review of neural networks in plant disease detection using hyperspectral data. *Inf. Process. Agric.*, 5, 3, 354–371, 2018.

58. de Vita, F., Nocera, G., Bruneo, D., Tomaselli, V., Giacalone, D., Das, S.K., Porting deep neural networks on the edge via dynamic K-means compression: A case study of plant disease detection. *Pervasive Mob. Comput.*, 75, 101437, 2021.
59. Tiwari, V., Joshi, R.C., Dutta, M.K., Dense convolutional neural networks based multiclass plant disease detection and classification using leaf images. *Ecol. Inform.*, 63, 101289, 2021.
60. Yogeshwari, M. and Thailambal, G., Automatic feature extraction and detection of plant leaf disease using GLCM features and convolutional neural networks. *Mater. Today: Proc.*, 2021.
61. Sathiya, V., Josephine, M.S., Jeyabalaraja, V., An automatic classification and early disease detection technique for herbs plant. *Comput. Electron. Agric.*, 100, 108026, 2022.
62. Ashwinkumar, S., Rajagopal, S., Manimaran, V., Jegajothi, B., Automated plant leaf disease detection and classification using optimal MobileNet based convolutional neural networks. *Mater. Today: Proc.*, 51, 480–487, 2021.
63. Al-windi, B.K.M.A. and Abbas, A.H., Eucalyptuses leaf disease detection using multiple analyses systems and backbrobagation neural network. *Mater. Today: Proc.*, 2021.
64. Wang, Y., Wang, H., Peng, Z., Rice diseases detection and classification using attention based neural network and bayesian optimization. *Expert Syst. Appl.*, 178, 114770, 2021.
65. Paymode, A.S. and Malode, V.B., Transfer learning for multi-crop leaf disease image classification using convolutional neural network VGG. *Artif. Intell. Agric.*, 6, 23–33, 2022.
66. Picon, A., Seitz, M., Alvarez-Gila, A., Mohnke, P., Ortiz-Barredo, A., Echazarra, J., Crop conditional convolutional neural networks for massive multi-crop plant disease classification over cell phone acquired images taken on real field conditions. *Comput. Electron. Agric.*, 167, 105093, 2019.
67. He, C., Li, X., Liu, Y., Yang, B., Wu, Z., Tan, S., Ye, D., Weng, H., Combining multicolor fluorescence imaging with multispectral reflectance imaging for rapid citrus Huanglongbing detection based on lightweight convolutional neural network using a handheld device. *Comput. Electron. Agric.*, 194, 106808, 2022.
68. Afonso, M., Blok, P.M., Polder, G., van der Wolf, J.M., Kamp, J., Blackleg detection in potato plants using convolutional neural networks. *IFAC-PapersOnLine*, 52, 30, 6–11, 2019.
69. Zhang, Y., Yu, J., Chen, Y., Yang, W., Zhang, W., He, Y., Real-time strawberry detection using deep neural networks on embedded system (rtsd-net): An edge AI application. *Comput. Electron. Agric.*, 192, 106586, 2022.
70. Zhang, W., Zhou, G., Chen, A., Hu, Y., Deep multi-scale dual-channel convolutional neural network for Internet of Things apple disease detection. *Comput. Electron. Agric.*, 194, 106749, 2022.

71. Mishra, S., Sachan, R., Rajpal, D., Deep convolutional neural network based detection system for real-time corn plant disease recognition. *Proc. Comput. Sci.*, 167, 2003–2010, 2020.
72. Yisa, M., Fadeyibi, A., Adisa, O., Finite element simulation of temperature variation in grain metal silo. *Res. Agric. Eng.*, 10, 3, 8–17, 2018.
73. Islam, M.P., Hatou, K., Aihara, T., Seno, S., Kirino, S., Okamoto, S., Performance prediction of tomato leaf disease by a series of parallel convolutional neural networks. *Smart Agric. Technol.*, 2, 100054, 2022.
74. Agarwal, M., Singh, A., Arjaria, S., Sinha, A., Gupta, S., ToLeD: Tomato leaf disease detection using convolution neural network. *Proc. Comput. Sci.*, 167, 293–301, 2020.
75. Hussein, M.A. and Abbas, A.H., Plant leaf disease detection using support vector machine. *Al-Mustansiriyah J. Sci.*, 30, 1, 105–110, 2019.
76. Rumpf, T., Mahlein, A.K., Steiner, U., Oerke, E.C., Dehne, H.W., Plümer, L., Early detection and classification of plant diseases with support vector machines based on hyperspectral reflectance. *Comput. Electron. Agric.*, 74, 1, 91–99, 2010.
77. Griffel, L.M., Delparte, D., Edwards, J., Using support vector machines classification to differentiate spectral signatures of potato plants infected with potato virus Y. *Comput. Electron. Agric.*, 153, 318–324, 2018.
78. Mokhtar, U., El-Bendary, N., Hassenian, A.E., Emary, E., Mahmoud, M.A., Hefny, H., Tolba, M.F., Mokhtar, U., Hassenian, A.E., Emary, E., Mahmoud, M.A., SVM-based detection of tomato leaves diseases. *Adv. Intell. Syst. Comput.*, 323, 641–652, 2015.
79. Rumpf, T., Mahlein, A.K., Steiner, U., Oerke, E.C., Dehne, H.W., Plümer, L., Early detection and classification of plant diseases with support vector machines based on hyperspectral reflectance. *Comput. Electron. Agric.*, 74, 1, 91–99, 2010.
80. Leena, N. and Saju, K.K., Classification of macronutrient deficiencies in maize plants using optimized multi class support vector machines. *Eng. Agric. Environ. Food*, 12, 1, 126–139, 2019.
81. Omrani, E., Khoshnevisan, B., Shamshirband, S., Saboohi, H., Anuar, N.B., Nasir, M.H.N.M., Potential of radial basis function-based support vector regression for apple disease detection. *Meas.: J. Int. Meas. Confed.*, 55, 512–519, 2014.

10

Biomimicry for Sustainable Structural Mimicking in Textile Industries

Mira Chares Subash and Muthiah Perumalsamy*

Department of Chemical Engineering, National Institute of Technology, Tiruchirappalli, India

Abstract

Biomimetics is the innovative technology, which incorporates the principles of engineering, physical science, chemical science, biological science for the manufacture of desired materials, clothing, functions, specifications, and applications. The employment of biomimetics in clothing industry improvise the future of biomimicry in textile industry. The biomimicry creates a bridge between the biological organism and future fabrics. Nature provokes the knowledge of structural designs, change in functionality, mitigate risk, cost reduction, materials, and sustainable energy. The advantages of textile biomimicry are self-repair, adhesion, energy conservation, and superhydrophobicity. The biomimicry beautifies the clothing generation by providing unexplored technical designs and functions.

Keywords: Biomimetics, textile, sustainable, future fabrics, energy conservation, natural resources

10.1 Introduction

The term "Biomimetics" is the combination of two Greek words "bios" and "mimesis" which refer to the meaning "life" and "to initiate," respectively [1]. The biomimicry creates a platform by copying, adaption, derivation from the natural resources. The study of the formation, structural information, biological product characteristics, and its mechanisms is called biomimetics. Biomimicry is a scientific term defined to mimic the elemental nature of the innovation in products. Textile biomimetics can be

**Corresponding author*: mpsamy@nitt.edu

Inamuddin, Tariq Altalhi and Ashjan Alrogi (eds.) Biomimicry Materials and Applications, (215–234)

generated by examining color combinations, patterns, and symmetrical objects. Biomimicry clothing aims to apply biological materials, microorganisms in fabric designs, and the production of new materials in textiles [2]. The advantages of mimicked fabric are sustainable, regenerative, energy preservation, and superhydrophobic. The commercial mimicked fabric is scarce which indulges the thirst for deep knowledge in the application of biomimetics in textile industries. The expression "textile" was procured from the Latin word "texere," which sense the meaning "to weave." The weaving process involves interlacement of assembled orthogonal long flexible strands or yarns. The examples of natural fibrous structures are *Cocus nucifera*, Linn, Coir fiber, wood, bamboo, and Leaf sheath [3]. The leaf sheath contains alternated fiber sheets oriented orthogonally to form a woven structure. The wood fiber consists of hemicellulose and lignin reinforced with hollow tubular cells. The helical structure of fiber contributes to the mechanical strength of the fiber. Bamboo is the strongest natural fiber composed of a hollow cylinder with equidistant nodes.

The first biomimicked fashion garments were introduced at the Sydney Olympic Games, in 2000. They represented swimsuits with shark's skin replication for the swimmer performance enhancement. Following the above, by the year 2007 Japanese Teijin company manufactured Morphotex fiber from the Morpho butterfly wings [4]. The morphotex fiber has inspired the audience with the vivid cobalt blue color even in the absence of pigments, performance similar to polyester and nylon fibers. The production process of Morphotex fibers was energy-saving and eco-friendly. Similarly, a specialist outdoor clothing company named Paramo has shown interest in adopting new materials leveraging biomimicry. The company has a keen thirst for innovative fabrics with quality performance in harsh conditions. The Paramo company employs the Nikwax, an innovative textile system for the production of highly qualified outdoor clothing in a wide price range.

10.2 Examples of Biomimicry Fabrics

10.2.1 Algae Fiber

Algae can exist as algal polymers, algal bends, algal biomass composite, and biomolecules. A German Israeli firm named Algalife manufactured fabrics and dyes from algae. The algal fabric is an eco-friendly approach with a zero-waste system. The schematic representation of algal fabric production is represented in Figure 10.1. The algal fabrics can be cultivated with the

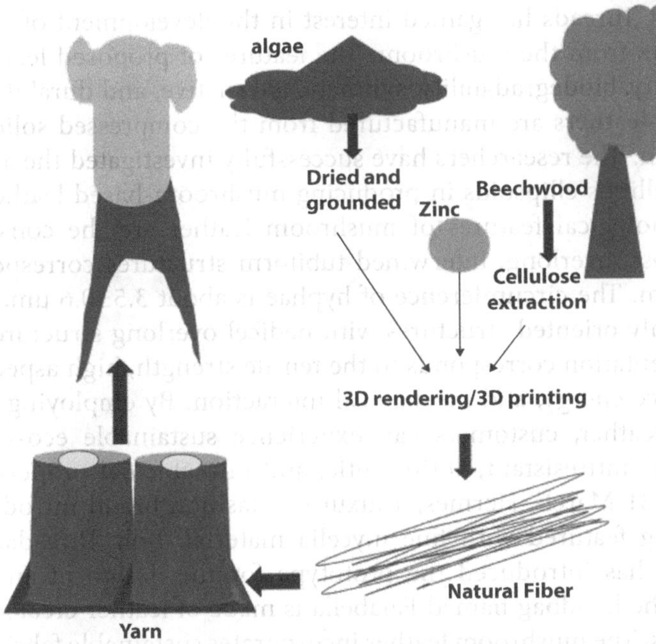

Figure 10.1 Schematic representation of algal fabrics production.

required water and sunlight. The commercialization of algal fabrics creates an environment with a reduction in pollution and energy consumption. The dyes employed in algal fabrics are non-allergic and chemical-free. Ulva of green algae and polymer (vinyl alcohol) were employed for the production of biocomposite plates by compression moulding and casting strategies. These plates were put to the test for rigidity and thermal degradation, the prominent execution was achieved with 30 wt.% of algal concentration [5]. The researchers in China have suggested that algal fabrics are heat resistant, non-retardants to toxic. Algae can beneficially replace plastic with biobased materials. Global bioplastic manufacturing has believed to reach 7.5 million by the year 2026.

10.2.2 Mushroom Leather

Plant-based leathers have the potential to revolutionize the fashion industry. A biotech company from California launched vegan leather derivated from fungal resources. Mushrooms consist of mycelium (root structure of mushrooms), which is rigid, densely intertwined, tough material with a network of thread-like structures. The mycelium is a fungal root-like structure composed of mass branching thread-like hyphae. A company named Bolt Threads has gained interest in the development of sustainable leathers from the mushroom. The features of proposed leather are sustainability, biodegradability, synthetic alternative, and durability. The mushroom leathers are manufactured from the compressed solid foam of mycelium. The researchers have successfully investigated the application of Phellinus ellipsoids in producing mushroom-based leathers [6]. The morphological features of mushroom leather are the constituent of criss-cross, overlong, interwined tubiform structures corresponding to mycelium. The circumference of hyphae is about 3.5±0.6 μm, which are randomly oriented structures with pedicel overlong structures. The hyphae orientation corresponds to the tensile strength, high aspect fiber ratio, surface energy, and mechanical interaction. By employing mushrooms as leather, customers can experience sustainable eco-friendly leather with antiresistant, antimycotic, and microbicidal properties. In the year 2021 March, Hermes, a luxurious fashion brand introduced a Victoria bag featured with fine mycelia material. Bolt Threads, Stella McCartney has introduced the prototype for the design of an iconic handbag. The handbag named Falabella is made of leather created from mushrooms. The mushroom leather incorporates sustainable fabrics with next-generation luxury materials. The synthesis of mushroom leather is schematically represented as Figure 10.2.

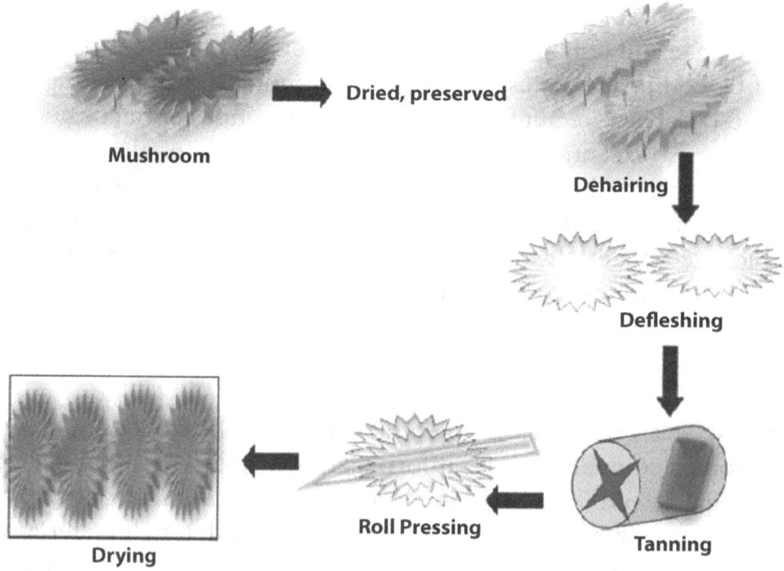

Figure 10.2 Schematic representation of mushroom leather.

10.2.3 Fabric Mimics

The fabric mimics are synthesized based on inspiration from nature. The Nikwax is a waterproof fabric that mimics animal fur. The fabric mimics are water repellent, and protectant against climatic conditions such as precipitation, condensation, evaporation, and heavy rain [7]. Insulating fabric requires three layers such as shell material, liner fabric, and sandwich layer. Based on the insulation method namely quilting, baffling the distribution of fabric in the garment appears. Primaloft is the brand which combines the features of fabric insulation with shells, liners into single material. The Primaloft fabric can be worn as standalone (next to skin fabric), durable with flexible comfort, and breathability. The hybrid nature of Primaloft fabric dyed with variant color spectrum, which tends to fresh feeling silhouettes, textures, and designs. The pioneer Primaloft fabric mimics which in turn leads to the formation of Primaloft NEXT, Primloft Evolve series. The Primaloft fabrics mimics the properties of animal fur by incorporating fibers with variant lengths and deniers. The advantage of the mentioned fabric is recyclability, lightweight, insulating gradient with single material, properties of animal fur, and sustainability.

10.2.4 Bacterial Pigments

Industries widely make use of pigments for the application of additives, color intensifiers, additives, synthetic dyes etc. Current awareness of human safety and environmental awareness has inflated the thirst for a non-toxic natural source of food colors. Textile industries utilize synthetic dyes and pigments for fabrication with color variants. Contrary to the conventional methods, a company in London named Faber Futures has employed bacteria for the synthesis of the required color. The bacteria can be subjected to a fermentation process for the creation of pigments which is unfadable over a long period [8]. Compared to conventional dye pigment production, bacterial pigments consume a very minimum quantity of water and are inferred to be sustainable. The following Table 10.1 represents the possible color pigments which can be extracted from the bacterial species [9].

10.2.5 Orange Fabrics

Global production of orange peel waste is tons seven hundred tons per year. The orange fabrics have been generated with the proposal of manufacturing sustainable fibers from citrus waste and byproducts of citrus

Table 10.1 Bacterial pigments from microbial species.

Bacterial species	Bacterial pigments	Color appearance
Agrobacterium aurantiacum sp. *Paracoccus caotinifaciens* sp. *Xanthophyllomyces dendrorhus* sp.	Astaxanthin	Pinkish ruby
Bradyrhizobium specie	Pigments of Canthaxanthin	Pitch black reddish
Flavobacterium specie *Paracoccus zeaxanthinifaciens* *Achromobacter* *Bacillus* *Brevibacterium specie* *Corynbacterium michigannise*	Pigments of Zeaxanthin	Amber Creamy Brown Orange yellow Greyish to creamish
Corynbacterium insidosum	Indigoiodine	Blue
Rugamonas ruba *Streptpoverticillium rubireticuli* *Vibrio gagogenes* *Altermonas ruba* *Rhodococcus maris* *Serratia marcescens* *Serratia rubideaea*	Prodigiosin	Cherry Bluish red
Staphylococcus aureus	Pigments of Staphyloxanthin	Blonde yellowish
Chromobacterium violaceum *Janthinobacterium lividum*	Pigments of Violacein	Mauve
Pseudomonas aeruginosa	Procyanidin	Bluish shaded green
Xanthomonas oryzae	Pigments of Xanthomonadin	Amber

juice [10]. The orange fiber intensifies the utilization of hundreds of thousands of orange peels as ethereal fabrics. The fabrics made from orange peels are soft, cuddly, easy to drape, and anti-wrinkle with a wide range and types. The citrus fabric mimics the features of citrus fruits with qualities of vitamin A, vitamin E, and vitamin C. People experience the release of vitamins and vitamin absorption through the skin by wearing orange fabrics. The company named "Orange Fiber fabrics" has been patented for successfully extracting silk-like cellulose from the orange waste. The students initially filed an Italian patent followed by the extension of the international protocol in the U.S, India, Brazil, the European Union, and Mexico. The "Global Change Award" was given for the potential and positive impact of orange fabrics. The company has created orange fabrics with the motto of "Salvatore Ferragamo" which implies the term "Responsible Passion." The orange fabrics are considered to be sustainable and innovative design. A fabric name "palazzo" was designed directly from mother nature's orange fabric without being hurdled as food waste. The orange peels are processed with a patented strategy for cellulose extraction and spinning for yarn manufacturing.

10.2.6 Protein Couture

The Japanese designer has influenced the application of brewed protein in couture preparation [11]. At the Paris fashion week event 2000, a Japanese designer presented brewed protein material as a potential source for a new couture collection. The protein coutures are modifiable and capable of diverse surface textures.

10.2.7 Natural Fiber Fabrics

The natural fiber can be either plant or animal fiber, such as silk, cashmere, cotton, jute, hemp, flax, coir, banana, etc. The tulle fiber is composed of extracellular proteoglycans released by the insect larvae. The mulberry silkworm derived from *Bombyx mori* is known as the best silk with a shimmering silk appearance [12]. The prism-like structure of silk cloth refracts incoming light at variant angles, which in turn produces different colors. The silk production occurs in Hymenoptera, silverfish, beeswax, wasps, ants, beetles, lacewings, fleas, midgets, and flies. The cashmere mohair, yarn, lambswool is formerly pronounced as cashmere fiber from cashmere goats. Cashmere wool can be used for making yarns, fabrics, sustainable textiles, and clothing materials over a period of hundreds of years. The cotton fiber is the purest form of cellulose obtained from seeds of Gossypium in the mallow family *Malvaceae.* The composition of cotton is 88% to 97% cellulose, waxes, proteins, and pectins. Cotton fibers are predominantly blended with nylon, polyester, linen, and wool for the best result of fiber [13]. The bast fiber is obtained from the stem of dicotyledonous plants. Examples of bast fibers are jute, flax, hemp, ramie, kenaf, kudzu, and mulberry. The bast fiber from Oak trees is the foremost woven fabric in the

Table 10.2 Fibers from biomaterial and their specifications.

Biomaterial	Application	Tensile strength (Gpa)	Elongation (%)	Modulus (Gpa)	Breaking energy (kJ kg–1)
Cocoon silk (Bombyx mori)	manufacture of cocoon	0.6	14	6	-
Dragline silk	production of dragline threads	0.7–2.3	22–39	9.5–3.0	130–915
Minor ampullate	dragline reinforcement	1	5	-	30
Flagelliform silk	synthesis of spiral web	0.1–0.5	300	1	100
Aciniform	envelop prey application	360	46	0.6	-
Aggregate	Sticky silk glue	-	517	-	-
Kevlar	Fiber	2.9–3.0	2.5–4.0	70–115	33
Nylon	Fiber	0.3–0.7	15–40	7–34	60

world. The current application of bast fibers is carpet, yarn, rope, geotextile, hessian or burlap, paper, sacks etc. The fabric material is located in the phloem, which is interconnected with the xylem and epidermis. Microbial retting or chemical degumming is done to separate the phloem from the xylem. Similarly, polygalacturonase enzymes are used for the degumming of bast fibers [14]. The following Table 10.2 represents the application of biomaterial in textiles and the specification of the corresponding fiber.

10.3 Fabric Production from Biomaterial

10.3.1 Soy Fabric

The term "soy silk fabric" is often proclaimed as vegan cashmere, soy milk, and soy fiber protein. The soy silk fabric is usually procured from the byproducts of the soy industry. The possible soy fabric thread pool variation ranges from a hundred to five hundred. The characteristic features of spy fabric are high breathability, low heat retention abilities, high stretchability, and is highly inclined to pilling and bubbling. The soy fabric was pioneered and manufactured in the United States. The top soy fabric exporters in the world are China and the United States of America. The applications of soy fabric are upholstery fabrics, sustainable dresses, tank tops, long skirts, coat suits, evergreen shirts, athletic textiles, bairn clothing, and bed linen with recommended washing temperatures. The synthesis of soy fabric from the soy silk protein is enlisted in the following flowchart. The process involved in soy fabric production are represented in Figure 10.3 as follows.

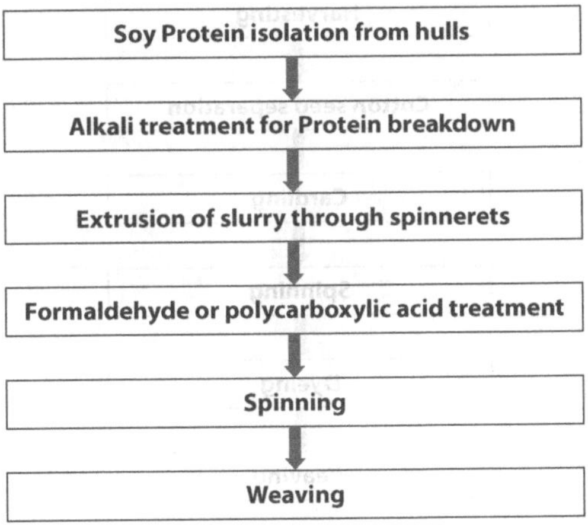

Figure 10.3 Schematic representation of soy fabric production.

10.3.2 Cotton Fabric

The term “Cotton Seed Fabric” is also known as “Pima cotton”, “Egyptian cotton” or “Supima cotton”. The extraction of cotton fabric is usually from the cottonseed plant. The fabric thread count which can be possibly extracted varies from 100 to 2000. The features of cotton fabric are high breathability, medium ardour retentiveness, medium whippiness, and extended susceptibility to bubbling and pilling. India is the large-scale creator and exporter of cotton in the world. The common applications of cotton are eco-friendly shirts, sustainable dresses, innovative design blouses, underwear, rigid socks, sweaters, sheets, blankets, design bags, profused jeans, and improvised skirts. The pioneer evidence of cotton fabric is from the Mehrgarh and Rakhigari parts of India during 5000 BC. Similarly, cotton cultivation flourished in the Indus Valley Civilization from 3300 to 1300 BC. The cotton fabric is well known for being adsorbent, quick-drying, high moisture-wicking, high heat, and easy draping.

10.3.3 Supima Fabric

Supima Fabric is obtained from the fibers of the *Gossypium barbadense* plant. The characteristic features of Supima fabric are height quality, softness, strength, color retention, luxury, moisture-wicking abilities, durability, breathability, lightweight, opaque, good fall, sheen, and staple fiber.

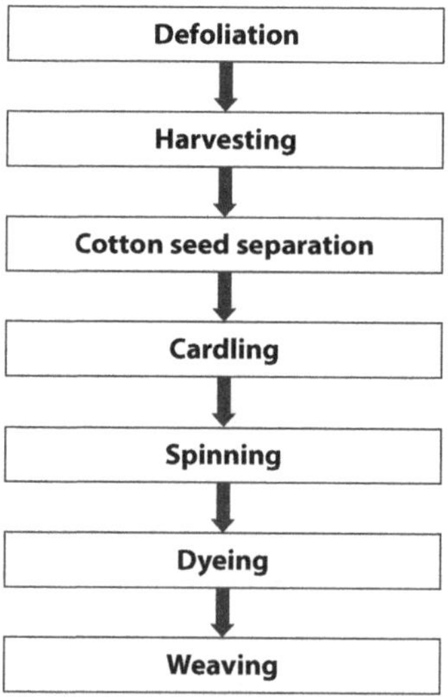

Figure 10.4 Schematic representation of cotton fabric production.

The Supima is the combination of cotton fibers and Pima fibers. The term Supima is derived from "superior" and "Pima". The inflation of Supima fabric happened in the year 1980s with 100,000 bales per year. The top five importers of Supima fabric are China, India, Pakistan, Turkey, and Peru. The Supima fabric is used for T-shirts, dress shirts, underwear, nightgowns, and pyjamas. The Supima cotton is genuinely certified by the ASA and USDA. The schematic representation of cotton fabric production is represented in Figure 10.4.

10.3.4 Pima Fabric

Pima fabric is produced from the specific type of cotton plant called *Gossypium barbadense* plant and is commonly referred to as "sea island cotton". The Pima is rare, high-quality cotton with predominant growth on the northern coast of Peru. The Pima fibers are 1.5x longer compared to regular cotton, and densely packed with 40mm length. The dense package of Pima fabric results in better dye absorption, lasting deep rich colors even after heavy wash. The Pima fibers can be cultivated naturally without using chemical pesticides, environmental risks and high recyclability and biodegradability. The commercial Pima fabric contains some polyester and lycra garments with major cultivation in the United States, Egyptian countries, and dry climate of Nile River Valley, West Indies since 3000 BC. One of the drawbacks of Pima fabric is its high susceptibility to frost damage. The incredible softness and rich texture of Pima fabric attract customers to its usage. One of the tedious processes of Pima fabric is seed picking which is done by the "cotton engine" during the mass production era.

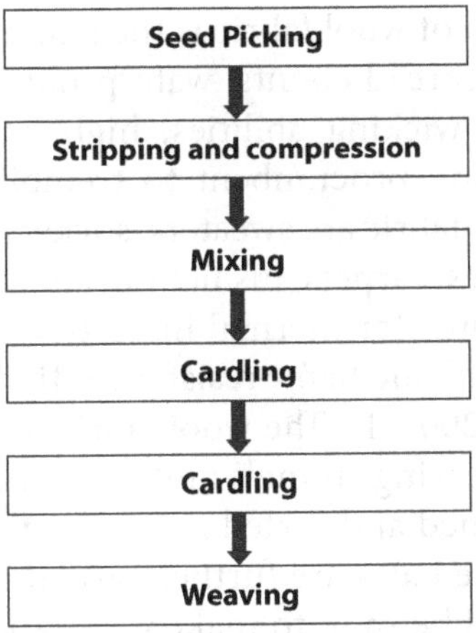

Figure 10.5 Schematic representation of spuima fabric production.

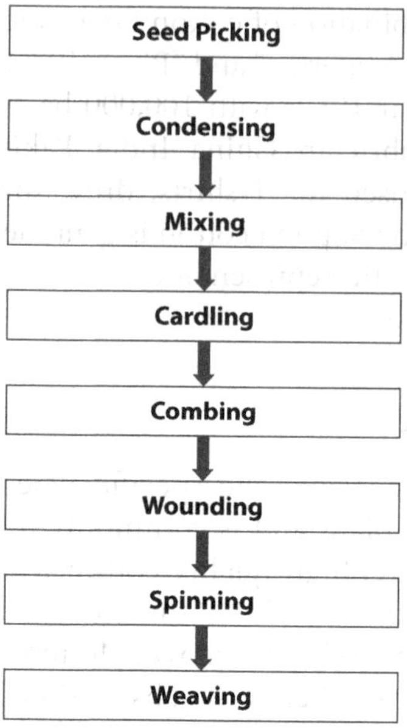

Figure 10.6 Schematic representation of pima fabric production.

The Pima cotton is significantly more expensive compared to another cotton. The schematic representation of spuima and pima fabric production is shown as Figures 10.5 and 10.6 respectively.

10.3.5 Wool Fabric

The common names of wool fabric are cashmere, fleece, fur, hair, and weed. The composition of wool fabric is the hair fibers from various animals, fragility, higher thread counts, waterproofness, absorbency, cushioning high moisture-wicking abilities, high heat retention, medium stretching, ductility, and procumbent to trembling and gurgling. The common usage of wool fabric are sweaters, socks, suits, pants, underwear, hats, gloves, gear clothes, carpets, insulation, and firefighting clothes. The wool fabric is well known for thermal insulation, bulkiness, and natural waviness namely "crimp" and flame resistance. The extraboliant wool production occurred by 1200 AD. The wool fabric production initiates with the shearing of wool-bearing animals multiple times throughout the year. The shorn wool is cleaned and sorted into bales with the employment of a chemical catalyst. The bales are further carded into long strand fibers. The carded strands will be spun to make yarns and the weaving process. The wool fabric production is schematically represented as Figure 10.7.

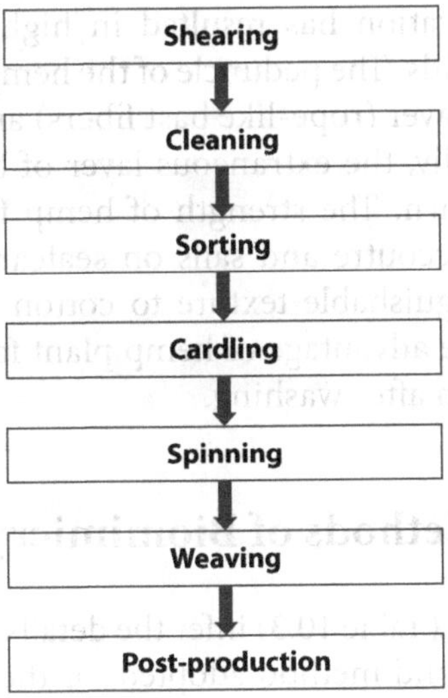

Figure 10.7 Schematic representation of wool fabric production.

10.3.6 Hemp Fabric

Industrial hemp is obtained from the peduncle of the *Cannabis sativa* plant with 250 to 300 possible garment count variations. The hemp fabric also possesses outsole, durability, high moisture-wicking abilities, low retention, and medium stretching and is less prone to pilling, mould-resistant, mildew, and bubbling. The applications of hemp fabric are hoodies, socks, apparel, T-shirts, dresses, and household textiles. The steps involved in the weaving of hemp fabric is schematically represented in Figure 10.8.

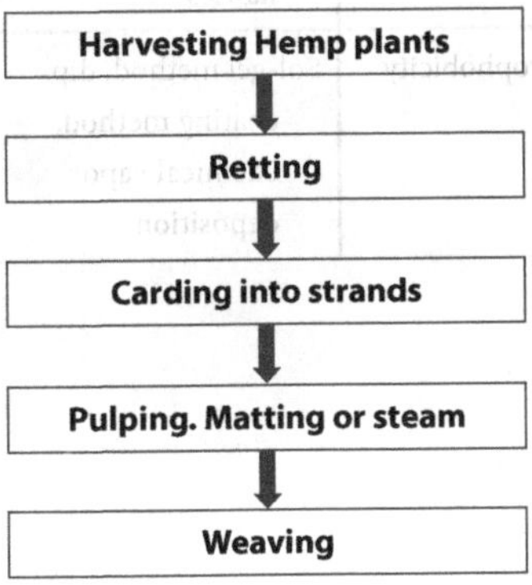

Figure 10.8 Weaving of hemp fabric.

The hemp fiber cultivation has resulted in high tetrahydrocannabinol (THC) and cannabinoids. The peduncle of the hemp plant contains couple of layers an exteriorr layer (rope-like bast fibers) and the intramural layer (woody pith). Generally, the extraneous layer of bast fibers is employed for making rope or yarn. The strength of hemp fiber makes people the foremost option for accoutre and sails on seafearing vessels. The hemp fabric has an indistinguishable texture to cotton but its thrice stronger than cotton fabric. The advantage of hemp plant fabric is its softness and non-degradability even after washing.

10.4 Current Methods of Biomimicry Materials

The following methods (Table 10.3) infer the details of various biomaterials and its characteristics and method adopted for the production of desired product.

Table 10.3 Current methods of biomimicry materials.

Biomaterials	Characteristics	Current method	Resultant product
Sharkskin	Antimicrobial effect	Nano/micro weaving with nano/microfilaments	Elastomers of polydimethylsiloxane
Sharkskin	Drag reduction	3D printing riblets	Macro riblets of knitted fabrics
Butterfly wing	Structural color	3D printed foams	Nylon, polyesters
Microstructure of the Fogstand battle	Fog harvesting	Synthetic monofilaments of 3D spaced knitted fabrics	Synthetic knitted fabrics
Nanostructure on Taro leaf	Hydrophobicity	Sol-gel method, dip coating method, chemical vapor deposition	Intricate micro/ nanostructures

10.5 Future of Biomimicry

Nature stands as an inevitable source of inspiration with massive technological inspiration. Most biomimicry materials are initiated based on people's demand for sustainable and environment-friendly textiles. Biomimicry expresses as the junction which combines science, creativity, design, and engineering [4]. The mimicked fabrics and biomaterial garments are biodegradable, effective protectants without compromising the customer's desired outcome. An online course named MOTIF's Sustainability in Fashion online course is being introduced with aim of developing the product with sustainability in fashion. The pioneer course connects the brand dealers, branching sources, vendors, and retailers in the fashion industry. The predominant advantage of the course is product lifecycle, sustainable strategies, and innovative design. The successful movement towards a circular fashion economy was accelerated during 2018 with commitments by leading fashion brands. The revolution in fashion design happened on March 11, 2020, by conducting MOTIF interaction sessions with clothing material designers, and researchers in the fashion industry towards the development of a future with sustainable fashion materials.

10.6 Benefits of Biomimicry

10.6.1 Sustainability

Biomimicry follows life principles such as a bottom-up approach, material optimization, free energy resources, cross-pollination, diversity embracement, adaption, and evolution of eco-friendly materials, symbiotic relationships, and biosphere enhancement.

10.6.2 Perform Welt

Biomimicry utilizes successful design strategies of the survivors, which result in better thriving in the marketplace. The design strategy is one of the crucial steps as nature will not be effective if the carrier dies. The vetting strategies have been employed by nature for over 3.8 billion years.

10.6.3 Energy Saving

Energy is the most expensive weapon in the human world. Therefore, energy conservation is crucial which can be accomplished by Biomimicry materials. In general, plants trap energy from sunlight in the hunt for resources, optimize usage of energy at every turn and deplete energy scarcity. By employing plant biomimicry material, energy can be efficiently translated with cost efficiency and great profitability.

10.6.4 Cut-Resistant Costs

The cut resistance cost denotes the making cost of the specified material. Nature has a distinct shape, and analysis of nature's structure and strategies. The biomimicry helps the manufacturers by minimizing the amount spent by the company, maximising the product effectiveness, design pattern improvisation, and achievement of target design and functions.

10.6.5 Eliminate Waste

Biomimicry creates a unique environment with nature transition materials, nutrient habitat, resource optimization, and waste elimination. The biomimicry materials organize the company's habitat flows more similar to nature's design, which results in profit, cost savings, creation of new profit centers with zero waste and positive feedback.

10.6.6 New Product Derivation

Biomimicry creates an innovation platform with a stale product vision in wide categories of radical light. The innovation platform builds disruptive technologies, transforms the industry, and builds entirely new industries. Similarly, biomimicry provokes new growth rates, customer attraction, innovation, and sustainability. Hence, the production of biomimicry materials reignites stale product categories, environment pro activation, and product innovation.

10.6.7 Disrupt Traditional Thinking

The introduction of biomimicry materials assists the manufacturer in exploring new solutions, creating of brainstorm opportunities, solving new challenges, and synthesis of innovative materials. The biomimicry disrupts traditional thinking by combining the best features of system thinking, design thinking, reversing, and testing engineer solutions.

10.6.8 Adaptability to Climate

Synthetic materials face major challenges in holding nature at bay, leveraging cyclic processes, building readily available materials and energy conservation. Biomimicry materials put a breakthrough in the sustainable goal achievement and adaptability to context and climate. The biomimicry material not only nourishes the system but also understands true sustainability from nature.

10.6.9 Nourish Curiosity

The biomimicry material provides the platform to satisfy customer curiosity and employment nature's genuine water, energy, and material design strategies. The perspective of biomimicry material broadens the solution space to nourish curiosity

10.6.10 Leverage Collaboration

The material inspired by nature identifies and adapts principles and recipes to change the key. The earth life is the epitome of resilience, adaption, changing context, the resilience of nature, diversity incorporation, redundancy, decentralization, and integration of rapid feedback loops. The biomimicry material creates human-built systems with resilient inheritance to change and disturbances. Further, the biomimicry process inherits a collaborative approach, cross-pollination, innovation, and organisational benefits. The products are designed with a rethinking of products, reimaging nature models, nature design mentoring, and cultivating collaborative relationships. The symbiotic relationships benefit the safety of resources, project cost, energy conservation, and sustainability.

10.7 Conclusion

Biomimicry products possess advantageous features namely the creation of a green economy, an increase in biodegradable products, energy conservation, and make a sustainable environment. In biomimetics, the structural design of systems and material production is modeled based on biological entities and processes. The theme of biomimicry is inspired by the translation of natural solution principles to human beings like copying a photosynthetic process and mimicking a nature-inspired building. The features of biomimicked textiles are functional surfaces, structural colors, self-healing, and insulation properties. The hurdles of biobased fabric production are slower agitation, size reduction during the shrinking process, and shorter cycles. Apart from the manufacturing hurdles, biomimetic fabrics are sustainable, readily available, have higher dye absorption, and energy conservation, and sustainable and eco-friendliness.

References

1. Das, S., Shanmugam, N., Kumar, A., Jose, S., Review: Potential of biomimicry in the field of textile technology. *Bioinspired, Biomim. Nanobiomaterials*, 6, 224–235, 2017, https://doi.org/10.1680/jbibn.16.00048.
2. Maglic, M.J., Biomimicry: Using nature as a model for design, pp. 1–88, 2014.
3. Chares Subash, M. and Muthiah, P., Eco-friendly degumming of natural fibers for textile applications: A comprehensive review. *Clean. Eng. Technol.*, 5, 100304, 2021, https://doi.org/10.1016/j.clet.2021.100304.
4. Chaurasia, M. and Srivastava, S., Biomimicry and its applications - a review. *Int. J. Eng. Appl. Sci. Technol.*, 04, 545–549, 2020, https://doi.org/10.33564/ijeast.2020.v04i12.098.
5. Constante, A., Pillay, S., Ning, H., Vaidya, U.K., Utilization of algae blooms as a source of natural fibers for biocomposite materials: Study of morphology and mechanical performance of Lyngbya fibers. *Algal Res.*, 12, 412–420, 2015, https://doi.org/10.1016/j.algal.2015.10.005.
6. Bustillos, J., Loganathan, A., Agrawal, R. *et al.*, Uncovering the mechanical, thermal, and chemical characteristics of biodegradable mushroom leather with intrinsic antifungal and antibacterial properties. *ACS Appl. Bio Mater.*, 3, 3145–3156, 2020, https://doi.org/10.1021/acsabm.0c00164.
7. Pegoretti, A., Fabbri, E., Migliaresi, C., Pilati, F., Intraply and interply hybrid composites based on E-glass and poly(vinyl alcohol) woven fabrics: Tensile and impact properties. *Polym. Int.*, 53, 1290–1297, 2004, https://doi.org/10.1002/pi.1514.
8. Duan, S., Feng, X., Cheng, L. *et al.*, Bio-degumming technology of jute bast by Pectobacterium sp. DCE-01. *AMB Express*, 6, 1, 86, 2016, https://doi.org/10.1186/s13568-016-0255-3.
9. Abdulkadir, N., Bacterial pigments and its significance. *MOJ Bioequiv. Bioavailab.*, 4, 285–288, 2017, https://doi.org/10.15406/mojbb.2017.04.00073.

10. Aly, S., Sayed, S.M., Zahran, M.K., One-step process for enzymatic desizing and bioscouring of cotton fabrics. *J. Nat. Fibers*, 7, 71–92, 2010, https://doi.org/10.1080/15440478.2010.481086.
11. Akinhanmi, T.F. and Atasie, V.N., Chemical composition and physicochemical properties of cashew nut. *J. Agric. Food Environ. Sci.*, 2, 1, 1–10, 2008.
12. Ayutsede, J., Gandhi, M., Sukigara, S. *et al.*, Regeneration of Bombyx mori silk by electrospinning. Part 3: Characterization of electrospun nonwoven mat. *Polym. (Guildf)*, 46, 1625–1634, 2005, https://doi.org/10.1016/j.polymer.2004.11.029.
13. Kohli, P. and Gupta, R., Alkaline pectinases: A review. *Biocatal. Agric. Biotechnol.*, 4, 3, 279–285, 2015, https://doi.org/10.1016/j.bcab.2015.07.001.
14. Wong, L.Y., Saad, W.Z., Mohamad, R., Tahir, P.M., Optimization of cultural conditions for polygalacturonase production by a newly isolated Aspergillus fumigatus R6 capable of retting kenaf. *Ind. Crops Prod.*, 97, 175–183, 2017, https://doi.org/10.1016/j.indcrop.2016.12.019.
15. Weerasinghe, D.U., Perera, S., Dissanayake, D.G.K., Application of biomimicry for sustainable functionalization of textiles: Review of current status and prospectus. *Text. Res. J.*, 89, 4282–4294, 2019, https://doi.org/10.1177/0040517518821911.

Index

Printed in the USA/Agawam, MA
September 16, 2026

910661.058